Mark Hannes Fischer

**Exotic electronic properties of ruthenates and related materials**

Mark Hannes Fischer

# Exotic electronic properties of ruthenates and related materials

effects of symmetry and orbital structure on the physics of the ruthenates

Südwestdeutscher Verlag für Hochschulschriften

**Impressum/Imprint (nur für Deutschland/only for Germany)**
Bibliografische Information der Deutschen Nationalbibliothek: Die Deutsche Nationalbibliothek verzeichnet diese Publikation in der Deutschen Nationalbibliografie; detaillierte bibliografische Daten sind im Internet über http://dnb.d-nb.de abrufbar.
Alle in diesem Buch genannten Marken und Produktnamen unterliegen warenzeichen-, marken- oder patentrechtlichem Schutz bzw. sind Warenzeichen oder eingetragene Warenzeichen der jeweiligen Inhaber. Die Wiedergabe von Marken, Produktnamen, Gebrauchsnamen, Handelsnamen, Warenbezeichnungen u.s.w. in diesem Werk berechtigt auch ohne besondere Kennzeichnung nicht zu der Annahme, dass solche Namen im Sinne der Warenzeichen- und Markenschutzgesetzgebung als frei zu betrachten wären und daher von jedermann benutzt werden dürften.

Verlag: Südwestdeutscher Verlag für Hochschulschriften GmbH & Co. KG
Dudweiler Landstr. 99, 66123 Saarbrücken, Deutschland
Telefon +49 681 37 20 271-1, Telefax +49 681 37 20 271-0
Email: info@svh-verlag.de

Approved by: Zürich, ETH, Diss., 2010

Herstellung in Deutschland:
Schaltungsdienst Lange o.H.G., Berlin
Books on Demand GmbH, Norderstedt
Reha GmbH, Saarbrücken
Amazon Distribution GmbH, Leipzig
**ISBN: 978-3-8381-2855-9**

**Imprint (only for USA, GB)**
Bibliographic information published by the Deutsche Nationalbibliothek: The Deutsche Nationalbibliothek lists this publication in the Deutsche Nationalbibliografie; detailed bibliographic data are available in the Internet at http://dnb.d-nb.de.
Any brand names and product names mentioned in this book are subject to trademark, brand or patent protection and are trademarks or registered trademarks of their respective holders. The use of brand names, product names, common names, trade names, product descriptions etc. even without a particular marking in this works is in no way to be construed to mean that such names may be regarded as unrestricted in respect of trademark and brand protection legislation and could thus be used by anyone.

Publisher: Südwestdeutscher Verlag für Hochschulschriften GmbH & Co. KG
Dudweiler Landstr. 99, 66123 Saarbrücken, Germany
Phone +49 681 37 20 271-1, Fax +49 681 37 20 271-0
Email: info@svh-verlag.de

Printed in the U.S.A.
Printed in the U.K. by (see last page)
**ISBN: 978-3-8381-2855-9**

Copyright © 2011 by the author and Südwestdeutscher Verlag für Hochschulschriften GmbH & Co. KG and licensors
All rights reserved. Saarbrücken 2011

to my parents

# Contents

| | |
|---|---|
| Contents | iii |
| List of Figures | vii |
| Acronyms | ix |

**1 Introduction**   1
- 1.1 The Ruddlesden-Popper Series of Layered Ruthenates . . . . . . .   1
- 1.2 Dimensionality, Magnetism and Criticality . . . . . . . . . . . .   4

**2 Dimensional Crossover in $Sr_2RuO_4$**   7
- 2.1 Introduction . . . . . . . . . . . . . . . . . . . . . . . . . . . .   7
- 2.2 Theoretical Description . . . . . . . . . . . . . . . . . . . . . .   10
  - 2.2.1 Slave-boson mean-field theory . . . . . . . . . . . . . .   11
  - 2.2.2 Inter-band hopping . . . . . . . . . . . . . . . . . . . .   13
  - 2.2.3 Single-electron Green's function . . . . . . . . . . . . .   14
- 2.3 Electronic Spectral Properties . . . . . . . . . . . . . . . . . . .   16
- 2.4 $c$-Axis Transport . . . . . . . . . . . . . . . . . . . . . . . . .   20
  - 2.4.1 Coherent transport . . . . . . . . . . . . . . . . . . . .   21
  - 2.4.2 Incoherent transport . . . . . . . . . . . . . . . . . . .   22
- 2.5 Conclusion . . . . . . . . . . . . . . . . . . . . . . . . . . . . .   24

**3 Metamagnetism and Nematicity in $Sr_3Ru_2O_7$**   27
- 3.1 Introduction . . . . . . . . . . . . . . . . . . . . . . . . . . . .   27
- 3.2 Microscopic Derivation of the Hamiltonian . . . . . . . . . . . .   32
  - 3.2.1 Basic hopping Hamiltonian . . . . . . . . . . . . . . . .   32

|  |  | 3.2.2 Magnetization and on-site interaction | 37 |
|---|---|---|---|
|  | 3.3 | Metamagnetic Transition | 41 |
|  |  | 3.3.1 Field in $z$ direction | 42 |
|  |  | 3.3.2 Field in $x$ direction | 43 |
|  |  | 3.3.3 Comparison | 46 |
|  | 3.4 | Nematic Phase | 49 |
|  | 3.5 | Discussion and Conclusion | 52 |
| **4** | **Superconductivity in Crystals with Locally Broken Inversion Symmetry** | | **57** |
|  | 4.1 | Introduction | 57 |
|  | 4.2 | Basic Theory of Superconductivity | 59 |
|  |  | 4.2.1 Green's functions and Gor'kov equations | 59 |
|  |  | 4.2.2 Crystal symmetry aspects | 61 |
|  |  | 4.2.3 Linearized gap equation | 63 |
|  | 4.3 | Inversion-Symmetry Lacking Bonds | 64 |
|  |  | 4.3.1 Symmetry classification | 64 |
|  |  | 4.3.2 Symmetry-allowed gap couplings | 66 |
|  |  | 4.3.3 Microscopic consideration of gap-function coupling | 67 |
|  |  | 4.3.4 Discussion | 71 |
|  | 4.4 | Inversion-Symmetry Lacking Layers | 72 |
|  |  | 4.4.1 Single layer | 73 |
|  |  | 4.4.2 Symmetry analysis of the stacked planes | 74 |
|  |  | 4.4.3 Gap couplings | 75 |
|  |  | 4.4.4 Discussion | 77 |
|  | 4.5 | Conclusions and Outlook | 79 |
| **A** | **SDW Instability** | | **83** |
| **B** | **Character Tables for $D_{4h}$** | | **87** |
| **C** | **Coupling of Different Gap Functions** | | **89** |
| **D** | **Structure of the Pairing Interaction** | | **91** |
| **E** | **Single Layer with Broken Inversion Symmetry** | | **95** |

| | |
|---|---:|
| **Bibliography** | **99** |

*CONTENTS*

# List of Figures

| | | |
|---|---|---|
| 1.1 | Ruddlesden-Popper series of layered ruthenates | 3 |
| 1.2 | Ruddlesden-Popper phase diagram | 5 |
| 2.1 | Experimental data on $Sr_2RuO_4$ | 9 |
| 2.2 | Dimensional crossover and possible hopping terms | 11 |
| 2.3 | Band structure as calculated from the spectral density | 17 |
| 2.4 | Spectral-density scans at constant momentum | 18 |
| 2.5 | Fermi surface of $Sr_2RuO_4$ | 19 |
| 2.6 | Quasiparticle weight at the Fermi energy | 20 |
| 2.7 | Resistivity in the coherent transport regime | 22 |
| 2.8 | Resistivity in the incoherent transport regime | 25 |
| 3.1 | Experiments on $Sr_3Ru_2O_7$ | 28 |
| 3.2 | Schematic phase diagrams of $Sr_3Ru_2O_7$ | 30 |
| 3.3 | Possible hoppings in the three-band model of $Sr_3Ru_2O_7$ | 33 |
| 3.4 | Single $RuO_4$ layer with rotated O octahedra | 34 |
| 3.5 | Real-space schematic of canted magnetization | 40 |
| 3.6 | Critical interaction strengths and density of states | 42 |
| 3.7 | Magnetization for $H \parallel c$ | 43 |
| 3.8 | Free energy analysis | 44 |
| 3.9 | Magnetization for $H \perp c$ | 46 |
| 3.10 | Magnetization jump and $T^*$ | 47 |
| 3.11 | Fermi-surface evolution for different field directions | 48 |
| 3.12 | Magnetization and nematic order parameter | 50 |
| 3.13 | Mean-field phase diagram with nematic phase | 51 |

LIST OF FIGURES

3.14 Dependence of nematic phase on forward-scattering strength and field direction . . . . . . . . . . . . . . . . . . . . . . . . . . . . 52
3.15 Width of nematic phase depending on forward-scattering strength 54
3.16 Anisotropic behavior and proximity to spin-density-wave instability 55

4.1 Inversion-symmetry lacking layers . . . . . . . . . . . . . . . . . . 73
4.2 Suppression of $T_c$ due to antisymmetric SOC . . . . . . . . . . . 78
4.3 Suppression of $T_c$ due to inter-layer hopping . . . . . . . . . . . 79

A.1 Analysis of spin-density-wave instability . . . . . . . . . . . . . 84

B.1 Stereographic projections for $D_{4h}$ and $C_{4v}$ . . . . . . . . . . . . 87

# Acronyms

| | |
|---|---|
| ARPES | angle-resolved photoemission spectroscopy |
| BZ | Brillouin zone |
| dHvA | de Haas-van Alphen |
| DOS | density of states |
| IR | irreducible representation |
| nn | nearest-neighbor |
| nnn | next-nearest-neighbor |
| QCEP | quantum critical endpoint |
| RPA | random-phase approximation |
| SDW | spin-density wave |
| SOC | spin-orbit coupling |
| TM | transition metal |
| vHS | van Hove singularity |

# Chapter 1

# Introduction

## 1.1 The Ruddlesden-Popper Series of Layered Ruthenates

The transition metal (TM) oxides are famous for displaying a plethora of electronic and magnetic properties as they range from metals and magnets to multiferroics and unconventional superconductors. These properties derive from their electronic structure with $d$ orbitals on the TM ions energetically close to the $p$ orbitals on the oxide ions, allowing for sizeable interactions between these two species. This can lead to rich phase diagrams showing ordered phases in the spin, charge, and orbital degrees of freedom. A very interesting family of TM compounds is the Ruddlesden-Popper series, which can be written with the general formula $A_{n-1}A'_2B_nX_{3n+1}$. In these crystal structures, the smaller cations B form layers of square lattices, each sitting in the middle of corner-sharing octahedral cages of the anions X [see Fig. 1.1(a)]. A unit cell of such a compound consists of $n$ $BX_4$ layers separated by the larger cations A and A'. A famous member of this series is $CaTiO_3$, the crystal that lends its name to the class of crystal structures known as perovskites.

Another very interesting class of materials realizing Ruddlesden-Popper phases, that has been the subject of intensive research, are the quasi-ternary ruthenium oxides $(Sr, Ca)_{n+1}Ru_nO_{3n+1}$. These layered ruthenates have attracted interest due to the diversity of occurring ground states including spin-triplet

## 1.1 The Ruddlesden-Popper Series of Layered Ruthenates

superconductivity [1], possibly an electronic nematic state [2], itinerant ferromagnetism [3, 4], antiferromagnetic Mott insulating behavior [5], and an orbital ordered state [6, 7, 8]. As these states are often in close competition, it is possible to tune between them with control parameters such as doping, pressure or magnetic fields. One such example is $Ca_{2-x}Sr_xRuO_4$, which, for $x < 0.2$, has an antiferromagnetic insulating ground state and exhibits metal-to-insulator transitions. Increasing the Sr doping to $0.2 \leq x < 0.5$, an antiferromagnetic metallic region can be observed merging to an almost ferromagnetic metal at $x_c \approx 0.5$. Finally, for $x = 2$ the system undergoes a superconducting transition below 1.5 K [5]. Within the wealth of phases and phenomena, we focus in this thesis on the strontium ruthenates $Sr_{n+1}Ru_nO_{3n+1}$ looking at some selected exotic electronic properties.

First prepared in 1959 [9], the infinite layer $SrRuO_3$ was found to have a ferromagnetic ground state with a Curie temperature of $T_C \approx 160$ K [3]. It is therefore a rare example of a $4d$ based ferromagnet and even the only known ferromagnetic metal among the $4d$ TM oxides, showing, however, unusual transport properties (being a so-called 'bad metal') [10]. Interestingly, the origin of ferromagnetism in $SrRuO_3$ is still not completely understood. On the other hand, the single layer $Sr_2RuO_4$ shows no sign of magnetic ordering and first appeared to be a realization of a strongly correlated two-dimensional Fermi liquid. $Sr_2RuO_4$ was thus initially used mainly as a perovskite substrate, especially for the high-temperature superconductors after their discovery in 1986 [11]. Looking for other materials with the layered structure of the high-$T_c$ superconductors, Maeno et al. found in 1994 that $Sr_2RuO_4$, which is isostructural to $La_2CuO_4$, the famous parent compound of the cuprate superconductors, became superconducting below 0.9 K [1]. The transition temperature is extremely sensitive to the sample purity and reaches approximately 1.5 K in the clean limit [12]. Unlike in the high-$T_c$ superconductors, the pairing of the superconducting state is likely to be in the triplet channel as suggested by NMR Knight Shift measurements [13]. In addition, $\mu$sR experiments [14] and Polar Kerr Rotation measurements [15] indicate a time-reversal-symmetry breaking superconducting state, leading to an analogue of the $p$-wave pairing in the A-phase of $^3$He [16]. A new aspect of chiral $p$ wave was introduced recently with the proposal of using the topologically stable, half-quantum vortices with a single Majorana zero mode for quantum computing [17].

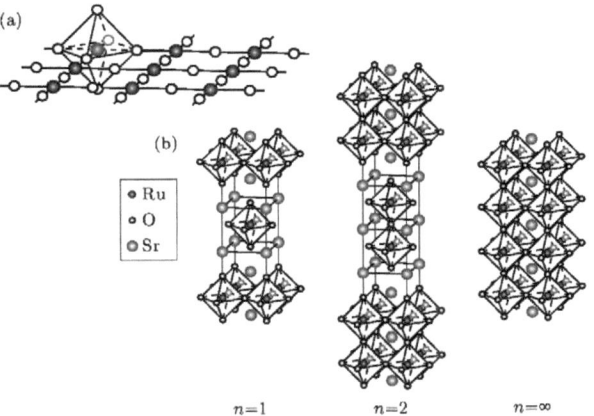

Figure 1.1: (a) The building block of the Ruddlesden-Popper phases consisting of corner-sharing oxygen octahedra with the Ru ions in the middle forming a square lattice. For clarity, only one octahedron is shown. (b) The quasi-one-dimensional single layer ($n = 1$), the intermediate bilayer ($n = 2$) and the three-dimensional ($n = \infty$) member of the Ruddlesden-Popper series of layered ruthenates.

The third compound of the Ruddlesden-Popper series of layered ruthenates that started to attract attention in recent years is the bilayer $Sr_3Ru_2O_7$. For sufficiently low temperatures, this material exhibits a metamagnetic transition, experimentally defined as a non-linear rise in the magnetization over a narrow region of applied magnetic field. The critical field crucially depends on the direction the field is applied and varies from $H_c^{xy} \approx 5.1\,\text{T}$ for in-plane fields to $H_c^z \approx 8\,\text{T}$ for fields parallel to the crystalline $c$ axis. An interesting feature is the appearance of additional structure in the phase diagram for out-of-plane fields with an unidentified low-temperature phase.

## 1.2 Dimensionality, Magnetism and Criticality

As the electronic structure in the Ruddlesden-Popper phases is dominated by the $RuO_4$ layers [Fig. 1.1(a)], these compounds show anisotropic transport depending on the number of adjacent layers in the unit cell. The resistivity anisotropy accordingly ranges from $\rho_c/\rho_{ab} \sim 10^3$ for $Sr_2RuO_4$ [18], over $\rho_c/\rho_{ab} \sim 300$ for $Sr_3Ru_2O_7$ [19] to three-dimensional behavior for $SrRuO_3$ [Fig. 1.1(b) shows the crystal structure of these three compounds]. Although $Sr_2RuO_4$ possesses Fermi-liquid behavior ($\rho = \rho_0 + AT^2$) in all directions at temperatures $T < 40\,K$, there is an extraordinary difference in the temperature dependence of the resistivity at higher temperature for in-plane and c-axis transport. While the in-plane resistivity is monotonically increasing, the c-axis resistivity shows a regime change from low-temperature metallic to high-temperature insulating behavior with a broad maximum in between at around $T^* \approx 130\,K$ [18]. This change is not associated with any structural transition or ordering phenomenon and has been attributed to a so far unspecified correlation effect of the electrons. In chapter 2, we discuss how this regime change can be understood in terms of a dimensional crossover within the basal plane.

Another property that is strongly influenced by the number of $RuO_4$ per unit cell is the tendency towards ferromagnetism. This dependence on the layer number is sketched in Fig. 1.2. While the quasi-two-dimensional $Sr_2RuO_4$ shows nice Fermi-liquid behavior and only a moderate enhancement in the ferromagnetic spin fluctuations is expected [20], the three dimensional $SrRuO_3$ is ferromagnetic. Evidence for the bilayer compound to be on the verge to ferromagnetism can be found in a relatively large Wilson ratio of approximately 10 [19] and a ferromagnetic transition under uniaxial pressure of about $1\,GPa$ [19]. Also, band-structure calculations point towards the same direction [21]. Early studies on $Sr_3Ru_2O_7$ even reported data indicating it to be ferromagnetic [22] which, however, is now believed was due to intergrowth with other phases. In addition, there is a metamagnetic transition as mentioned above. A very intriguing feature is that, while it is only a crossover for higher temperatures, for temperatures below $T^* \approx 1.2\,K$ the metamagnetic transition becomes first order for in-plane fields. When the field is tilted out of plane, this critical temperature $T^*$ decreases until it vanishes at approximately $80°$. The field angle can therefore be used to tune the critical

*Introduction*

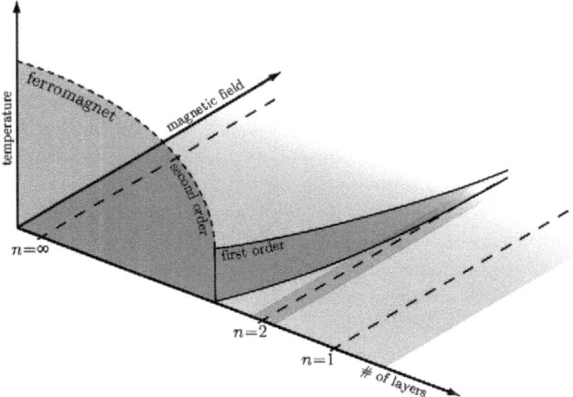

Figure 1.2: Phase diagram for the Ruddlesden-Popper series where the layer number is used as a control parameter to tune the compound's proximity to ferromagnetism. For the bilayer $Sr_3Ru_2O_7$, the close proximity leads to a metamagnetic transition for fields between 5 T and 8 T depending on the field direction. The darker area indicates the possibility of using the field direction for fine tuning in this compound.

endpoint to zero, thus realizing a quantum critical endpoint (QCEP) [23, 24]. This kind of control parameter is experimentally more easily accessible than pressure or chemical doping and thus, $Sr_3Ru_2O_7$ allows for the detailed study of quantum criticality. When trying to reach the putative QCEP in $Sr_3Ru_2O_7$ with ultra-high-quality single crystals, Grigera *et al.* [25] found a new phase forming below 1 K hiding the quantum critical endpoint. This intermediate phase, that exists in a small, muffin-shaped region in the $T$-$H$ plane, was shown to exhibit an in-plane-transport anisotropy under special field conditions and was therefore labelled an 'electronic nematic phase' [2]. Interestingly, this phase only appears for fields applied close to the *c*-axis direction. Chapter 3 addresses the question why the intermediate phase enclosing the QCEP in $Sr_3Ru_2O_7$ only appears for

## 1.2 Dimensionality, Magnetism and Criticality

fields applied almost parallel to the $c$ axis. For this purpose, we investigate the effect of a lattice distortion found in the bilayer compound, where the oxygen octahedra are rotated around the $c$ axis in opposite directions for neighboring octahedra. This rotation shifts the oxygen between Ru sites into off-center positions leading to a staggered spin-orbit coupling (SOC) of Rashba type. Such a coupling naturally leads to an anisotropic response of the system to an external magnetic field and we show how this can explain the anisotropy in the appearance of the intermediate phase.

The staggered SOC obtained for $Sr_3Ru_2O_7$ is reminiscent of the antisymmetric SOC found in crystals without inversion symmetry. These systems are of interest in the study of possible superconducting states as the lack of inversion symmetry was long believed to strongly suppress any form of spin-triplet pairing [26]. Indeed, the staggered SOC follows from a symmetry reduction on the Ru bonds, however, the global inversion symmetry of the system is retained. In chapter 4, we discuss possible ways of periodically breaking the inversion symmetry in globally inversion symmetric crystals ('antiferro-type ordering') and examine the associated effect on the superconducting instabilities and possible gap mixing. For this purpose, we analyze two specific examples and also discuss their connection to superconductivity in antiferromagnets, i.e., in systems with a 'staggered' time-reversal-symmetry breaking.

# Chapter 2

# Dimensional Crossover in $Sr_2RuO_4$

> *The dimensional crossover from the quasi-one-dimensional $d_{zx}$, $d_{yz}$ bands to a two-dimensional system due to a weak hybridization between these perpendicular bands is studied in this chapter. The corresponding two-orbital Hubbard model is treated within a slave-boson mean-field theory to take correlation effects such as the spin-charge separation in the one-dimensional bands into account. Using an RPA-like formulation for the Green's function of collective spinon-holon excitations the emergence of quasiparticles for low temperatures is explored. The results are used to discuss the evolution of the spectral density and the c-axis transport within a tunneling approach. For the latter, a change between a low- and high-temperature regime is found in qualitative accordance with experimental data.*

## 2.1 Introduction

Physical effects due to reduced dimensionality are most prominently visible for strongly correlated electrons. While under generic conditions low-energy electronic excitations retain their quasiparticle character in three and most likely two dimensions, correlated electrons in one dimension fractionalize in separate

## 2.1 Introduction

collective charge and spin excitations. All solids are three dimensional and low dimensionality appears through specific highly anisotropic electronic structure. In many cases, the effective dimensionality of such systems depends on the difference of energy scales which can lead to dimensional crossovers as systems are cooled or system parameters are changed, for example by applying pressure. The change of effective dimensionality is often marked through the onset of ordered states which are suppressed in low-dimensional systems due to thermal or quantum fluctuations. Such changes are well known from organic compounds such as $(TMTSF)_2X$, which undergoes a sequence of transitions as function of temperature, pressure, and chemical composition (X) [27]. Another remarkable feature is the modification of transport properties, which can depend on the nature of the charge carriers or on a change from coherent to incoherent transport. We argue in this chapter that $Sr_2RuO_4$, the quasi-two-dimensional member of the Ruddlesden-Popper series, is an example of such a system. As discussed in the introduction, the $c$-axis resistivity exhibits a crossover from Fermi-liquid behavior for low temperatures to insulating behavior for high temperatures. Figure 2.1(a) also shows the monotonic increase of the in-plane resistivity upon increasing temperature. The large resistivity anisotropy suggests a description of the electronic structure starting from planes with two-dimensional electron systems stacked in the $c$ direction. The rather strong renormalization of both, the effective mass and susceptibility of $Sr_2RuO_4$ further suggests that these systems should be treated as strongly correlated Fermi liquids. For the electrons to tunnel between layers, sufficient quasiparticle weight and lifetime in the planes, respectively, are required. A rapid loss of quasiparticle weight upon increasing the temperature was observed in angle-resolved photoemission spectroscopy (ARPES) measurements [28] and attributed to scattering processes. Motivated by this behavior we propose here a mechanism based on a dimensional crossover within the basal plane of $Sr_2RuO_4$ which affects the interlayer transport. Alternative explanations based on electron-phonon coupling [29, 30] and charge fluctuations caused by the tunneling electrons [31] have been discussed in the literature.

$Sr_2RuO_4$ is one of the best studied Fermi liquids with a Fermi surface measured with several techniques such as de Haas-van Alphen (dHvA) [34], Shubnikov-de Haas [35], ARPES [33] and Compton scattering [36]. All these measurements are in good agreement with band-structure calculations [37, 38] and show three

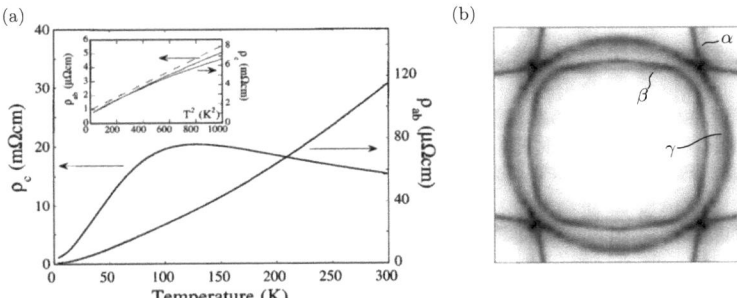

Figure 2.1: (a) The anisotropic resistivity of $Sr_2RuO_4$ as measured by Hussey and collaborators [32]. While the in-plane resistivity continues to grow monotonically, the c-axis transport shows a crossover to an insulating behavior. The inset is for comparison of the data with a $T^2$ dependence (dashed line) for low temperatures showing a nice Fermi-liquid behavior for all directions. (b) ARPES measurements showing the two quasi-one-dimensional $\alpha$ and $\beta$ bands as well as the two-dimensional $\gamma$ band crossing the Fermi energy. This graph is taken from Ref. [33].

bands crossing the Fermi energy with an average filling of 4/3 [see Fig. 2.1(b)]. Their shape is well accounted for by considering a tight-binding model of a single $RuO_4$ plane with Ru sites (4d orbitals) and oxygen sites (2p orbitals). Due to the oxygen octahedra surrounding the Ru atoms, the 4d orbitals are ligand-field split into $t_{2g}$ and $e_g$ orbitals with the latter ones only contributing to bands lying above the Fermi energy. The three bands that need to be considered thus stem from the three $t_{2g}$ orbitals $\pi$ hybridizing with the O 2p orbitals. The $d_{xy}$ orbital lies in the $RuO_2$ plane and forms the genuinely two-dimensional $\gamma$ band by hybridizing with the oxygen $p_x$ and $p_y$ orbitals. The other two orbitals $d_{yz}$ and $d_{zx}$, forming the so-called $\alpha$ and $\beta$ bands, do not hybridize with the $\gamma$ band as they have different parity with respect to reflection at the basal plane [The parity of the orbitals forming the $\alpha$ and $\beta$ bands is indicated in Fig. 2.2(b)]. In contrast to the $\gamma$ band which has practically no interlayer coupling, these other

## 2.2 Theoretical Description

two orbitals are also responsible for the weak interlayer hybridization [34]. Thus, in order to understand the c-axis transport, we focus our attention to the $d_{yz}$ and $d_{zx}$ orbitals which have a peculiar hybridization topology within the basal plane. Both orbitals form - through nearest-neighbor (nn) intra-orbital $\pi$-hybridization - an essentially one-dimensional band dispersion along the $y$ direction ($x$ direction) for the $d_{yz}$ orbital ($d_{zx}$ orbital). As shown in Fig. 2.2(b), the inter-orbital coupling occurs through considerably weaker next-nearest-neighbor (nnn) hybridization between $d_{yz}$ and $d_{zx}$ and leads to two two-dimensional bands which nevertheless keep strong one-dimensional features [see Fig. 2.2(a)] with pronounced nesting properties reflected in enhanced incommensurate magnetic spin fluctuations [39]. We argue in the following that the structure of these orbitals could lead to a crossover behavior between the one- and two-dimensional regime at the energy scale associated with the next-nearest-neighbor inter-orbital coupling similar to other systems discussed in the literature [27]. For this purpose, we develop a slave-boson scheme on the model of the $d_{yz}$-$d_{zx}$ orbitals in order to describe the fractionalization of the quasiparticles in the one-dimensional systems. We show that this method allows us to qualitatively account for the emergence of quasiparticle states in the hybridized bands in a rather simple manner. This method is also used to give a qualitative discussion of the regime change in the interlayer transport which we consider in the tunneling limit.

## 2.2 Theoretical Description

As mentioned above, we restrict our analysis to the bands originating from the Ru $d_{zx}$ and $d_{yz}$ as well as the O $p_z$ orbitals. After integrating out the oxygen degrees of freedom, we are to first order left with two independent one-dimensional bands, henceforth labeled $\nu = 1, 2$ corresponding to the $x$ and $y$ directions of the $d_{zx}$ and $d_{yz}$ orbitals, respectively. For the description of these two bands we use a two-band Hubbard Hamiltonian,

$$\mathcal{H} = -t \sum_{\nu,s} \sum_{\langle i,j \rangle_\nu} c^\dagger_{\nu is} c_{\nu js} + U \sum_{\nu,i} n_{\nu i\uparrow} n_{\nu i\downarrow} \tag{2.1}$$

with $t$ the hopping integral, $U$ the on-site Coulomb interaction and $\langle i, j \rangle_\nu$ denoting nearest neighbors in the corresponding direction $\nu$. The operator $c^\dagger_{\nu is}$ creates

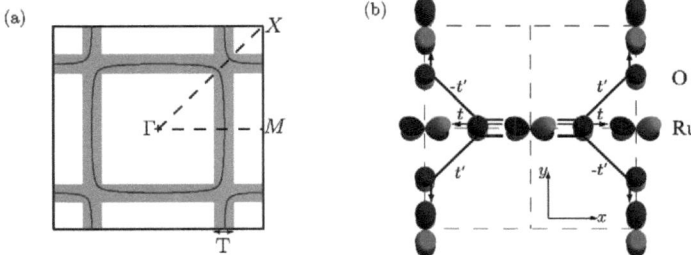

Figure 2.2: (a) The two-dimensional nature of the Fermi surfaces of the bands stemming from the $d_{yz}$ and $d_{zx}$ orbitals is lost for temperatures above the weak inter-orbital hybridization. In addition, the dashed line shows the path along which Fig. 2.3 is plotted. (b) The hopping structure for the $d_{zx}$ and $d_{yz}$ electrons (here for a $d_{zx}$ placed in the middle): in lowest order only via an O $2p$ to the same orbital of a nearest neighbor or as a next-order term via 2 O $2p$ to the other orbital of a next-nearest neighbor. To indicate the orbitals' parity with respect to reflection at the basal plane, the orbitals are drawn slightly tilted.

an electron in the orbital $\nu$ at site $i$ with spin $s$ and $n_{\nu i s} = c^\dagger_{\nu i s} c_{\nu i s}$. Due to the small overlap of the Wannier functions for the different bands, only an intra-band on-site interaction is included. Before continuing by introducing a term hybridizing these two bands, we want to deal with the strong correlations first.

### 2.2.1 Slave-boson mean-field theory

Since such correlated electron systems are difficult to handle we seek an approximation by assuming that Coulomb repulsion is sufficiently strong as to essentially suppress double occupancy, leading to the non-holonomic constraint

$$\sum_s c^\dagger_{\nu i s} c_{\nu i s} \leq 1. \qquad (2.2)$$

We may use a slave-boson technique which has been developed in the context of the $t$-$J$ model to simplify this constraint. By writing the electron operators

## 2.2 Theoretical Description

$c_{\nu i s}$ in terms of new fermionic operators representing the spin degrees of freedom (spinons) and bosonic operators for the charge degrees of freedom (holons), $c_{\nu i s} = f_{\nu i s} b^{\dagger}_{\nu i}$, and stipulating that each site is either occupied by a holon or a spinon,

$$b^{\dagger}_{\nu i} b_{\nu i} + \sum_s f^{\dagger}_{\nu i s} f_{\nu i s} = 1, \qquad (2.3)$$

double occupancy is strictly inhibited [40]. Note that this decomposition conserves the anticommutation relation of the electron operators $c_{\nu i s}$. From this replacement, the so-called slave-boson Hamiltonian results,

$$\begin{aligned}\mathcal{H} = & -t \sum_{\nu,s} \sum_{\langle i,j \rangle_\nu} f^{\dagger}_{\nu i s} b_{\nu i} b^{\dagger}_{\nu j} f_{\nu j s} \\ & + \sum_{\nu,i} \lambda_{\nu i} \left( b^{\dagger}_{\nu i} b_{\nu i} + \sum_s f^{\dagger}_{\nu i s} f_{\nu i s} - 1 \right) \\ & - \mu \sum_{\nu,i,s} f^{\dagger}_{\nu i s} f_{\nu i s} \end{aligned} \qquad (2.4)$$

with Lagrange multipliers $\lambda_{\nu i}$ enforcing the (local) constraints of one spinon or holon per site given in Eq. (2.3). In addition, we have introduced a chemical potential for the spinons to later fix the spinon concentration. To analyze this Hamiltonian we resort to the following standard approximations: We decouple the first term by introducing mean fields for the coherent motion of holons and spinons as

$$\chi^b_{\nu i} = \langle b^{\dagger}_{\nu i} b_{\nu j} \rangle \qquad (2.5)$$

and

$$\chi^f_{\nu i} = \sum_s \langle f^{\dagger}_{\nu i s} f_{\nu j s} \rangle, \qquad (2.6)$$

respectively, with $i,j$ being nearest neighbors. The treatment of the Hamiltonian is then further simplified by taking the mean fields as being independent of band and site and replacing the local constraints by a global one, i.e., $(\chi^b_{\nu i}, \chi^f_{\nu i}, \lambda_{\nu i}) \to (\chi^b, \chi^f, \lambda)$.

With these approximations, we arrive at a slave-boson mean-field Hamiltonian representing two independent species of particles, spinons and holons, in two independent one-dimensional bands. Changing to the momentum-space representation of the operators,

$$f^\dagger_{\nu j s} = \frac{1}{N}\sum_{\mathbf{k}} e^{-i\mathbf{k}\cdot\mathbf{r}_j} f^\dagger_{\nu\mathbf{k}s}, \tag{2.7}$$

$$b_{\nu j} = \frac{1}{N}\sum_{\mathbf{k}} e^{i\mathbf{k}\cdot\mathbf{r}_j} b_{\nu\mathbf{k}} \tag{2.8}$$

with $N$ the number of sites in the $x$ or $y$ direction, this Hamiltonian assumes the simple form

$$\mathcal{H}_0 = \sum_{\nu,\mathbf{k},s} \varepsilon_{\nu\mathbf{k}} f^\dagger_{\nu\mathbf{k}s} f_{\nu\mathbf{k}s} + \sum_{\nu,\mathbf{k}} \omega_{\nu\mathbf{k}} b^\dagger_{\nu\mathbf{k}} b_{\nu\mathbf{k}} \tag{2.9}$$

with the spinon energy $\varepsilon_{\nu\mathbf{k}} = -t\chi^b \gamma_{k_\nu} + \lambda - \mu$, the holon energy $\omega_{\nu\mathbf{k}} = -t\chi^f \gamma_{k_\nu} + \lambda$ and $\gamma_{k_\nu} = 2\cos k_\nu$. In this way, we effectively incorporate the spin-charge separation of the one-dimensional correlated electron systems realized in the nearest-neighbor-hopping Hamiltonian (2.1). The corresponding self-consistency equations following from Eqs. (2.5) and (2.6) read

$$\begin{aligned}
\chi^b &= \frac{1}{2N}\sum_{k_\nu}\frac{\gamma_{k_\nu}}{e^{\beta\omega_{\nu k}}-1}, \\
\chi^f &= \frac{1}{2N}\sum_{k_\nu,s}\frac{\gamma_{k_\nu}}{e^{\beta\varepsilon_{\nu k}}+1}, \\
x &= \frac{1}{N}\sum_{k_\nu}\frac{1}{e^{\beta\omega_{\nu k}}-1}, \\
1-x &= \frac{1}{N}\sum_{k_\nu,s}\frac{1}{e^{\beta\varepsilon_{\nu k}}+1},
\end{aligned} \tag{2.10}$$

where $\beta = T^{-1}$ is the inverse temperature ($k_B = 1$) and $x$ is the average holon number per site. Fixing $x$ thus allows for a self-consistent determination of the values of $\lambda$, $\mu$, $\chi^b$ and $\chi^f$.

## 2.2.2 Inter-band hopping

Next, we add the nnn inter-band hopping via two O $2p$ orbitals as displayed in Fig. 2.2(b). This weaker hopping term connects the two one-dimensional electron

## 2.2 Theoretical Description

systems and may be written in terms of the spinon and holon operators as

$$\mathcal{H}' = \sum_{s} \sum_{\mathbf{k},\mathbf{k}',\mathbf{q}} \left( g_{\mathbf{q}} f^{\dagger}_{1\mathbf{k}+\mathbf{q}s} b_{1\mathbf{k}} b^{\dagger}_{2\mathbf{k}'} f_{2\mathbf{k}'+\mathbf{q}s} + \text{h.c.} \right) \quad (2.11)$$

with the hopping structure factor $g_{\mathbf{q}} = -4t' \sin q_x \sin q_y$ and the inter-band hopping integral $t'$. We now use the fact that only physical particles, namely electrons, can be transferred between the two systems, invoking that a spinon and a holon have to correlate to yield hopping. Thus, we may interpret $\mathcal{H}'$ as an effective interaction term introducing attractive coupling between the two subspecies with the tendency to recombine them to an electronic quasiparticle [41].

### 2.2.3 Single-electron Green's function

To investigate the electronic spectrum in the following, we calculate the retarded single-electron Green's function defined as

$$G^{\nu\nu'}(\mathbf{q},t) = -\Theta(t) \langle \{ c_{\nu\mathbf{q}s}(t), c^{\dagger}_{\nu'\mathbf{q}s}(0) \} \rangle \quad (2.12)$$

which is independent of spin. Here, $\langle . \rangle$ denotes the expectation value with respect to the Hamiltonian $\mathcal{H} = \mathcal{H}_0 + \mathcal{H}'$, $\Theta(t)$ is the Heaviside step function and the operators are given in the Heisenberg picture,

$$c^{\dagger}_{\nu\mathbf{q}s}(t) = e^{i\mathcal{H}t} c^{\dagger}_{\nu\mathbf{q}s} e^{-i\mathcal{H}t}. \quad (2.13)$$

In Eq. (2.13), the electron operators are then replaced by the spinon and holon operators,

$$c^{\dagger}_{\nu\mathbf{q}s} = \frac{1}{N} \sum_{\mathbf{k}} f^{\dagger}_{\nu\mathbf{k}+\mathbf{q}s} b_{\nu\mathbf{k}}. \quad (2.14)$$

Following standard methods we first introduce an auxiliary Green's function

$$g^{\nu\nu'}_{\mathbf{k}}(\mathbf{q},t) = -\Theta(t) \langle \{ b^{\dagger}_{\nu\mathbf{k}}(t) f_{\nu\mathbf{k}+\mathbf{q}s}(t), c^{\dagger}_{\nu'\mathbf{q}s}(0) \} \rangle \quad (2.15)$$

which, in energy space, we write as

$$g^{\nu\nu'}_{\mathbf{k}}(\mathbf{q},E) = \int dt\, e^{-iEt} g^{\nu\nu'}_{\mathbf{k}}(\mathbf{q},t) \equiv \langle\langle b^{\dagger}_{\nu\mathbf{k}} f_{\nu\mathbf{k}+\mathbf{q}s}; c^{\dagger}_{\nu'\mathbf{q}s} \rangle\rangle_E. \quad (2.16)$$

This function can be calculated using the equation of motion,

$$Eg_\mathbf{k}^{\nu\nu'}(\mathbf{q}, E) = \frac{1}{N}\sum_\mathbf{p} \left\langle \left\{ b_{\nu\mathbf{k}}^\dagger f_{\nu\mathbf{k+q}s}, f_{\nu'\mathbf{p+q}s}^\dagger b_{\nu'\mathbf{p}} \right\} \right\rangle$$
$$- \left\langle\!\left\langle \left[ \mathcal{H}, b_{\nu\mathbf{k}}^\dagger f_{\nu\mathbf{k+q}s} \right]; c_{\nu'\mathbf{q}s}^\dagger \right\rangle\!\right\rangle_E. \quad (2.17)$$

This equation of motion involves commutators of the form

$$\left[ \mathcal{H}_0, b_{\nu\mathbf{k}}^\dagger f_{\nu\mathbf{k+q}s} \right] = -E_{\mathbf{k+q,k}}^{\nu\nu} b_{\nu\mathbf{k}}^\dagger f_{\nu\mathbf{k+q}s} \quad (2.18)$$

with $E_{\mathbf{k+q,k}}^{\nu\nu} = \varepsilon_{\nu\mathbf{k+q}} - \omega_{\nu\mathbf{k}}$ and

$$\left[ \mathcal{H}', b_{\nu\mathbf{k}}^\dagger f_{\nu\mathbf{k+q}s} \right] = \sum_{\mathbf{k',q'},s'} g_{\mathbf{q}'} f_{\nu\mathbf{k+q}'s'}^\dagger f_{\nu\mathbf{k+q}s} b_{\bar\nu\mathbf{k}'}^\dagger f_{\bar\nu\mathbf{k'+q}'s'}$$
$$+ \sum_{\mathbf{k',q'}} g_{\mathbf{q}'} b_{\nu\mathbf{k}}^\dagger b_{\nu\mathbf{k+q-q}'} b_{\bar\nu\mathbf{k}'}^\dagger f_{\bar\nu\mathbf{k'+q}'s}, \quad (2.19)$$

where for the index $\nu$ we define $\bar 1 = 2$ and vice versa. The higher order Green's functions coming from Eq. (2.19) are treated within random-phase approximation (RPA) to yield

$$\hat G(\mathbf{q}, E) = \hat G_0(\mathbf{q}, E) + \hat G_0(\mathbf{q}, E) \hat g(\mathbf{q}) \hat G(\mathbf{q}, E), \quad (2.20)$$

where

$$\hat G_0(\mathbf{q}, E) = \begin{pmatrix} G_0^1(\mathbf{q}, E) & 0 \\ 0 & G_0^2(\mathbf{q}, E) \end{pmatrix}, \quad (2.21)$$

$$\hat g(\mathbf{q}) = \begin{pmatrix} 0 & g_\mathbf{q} \\ g_\mathbf{q} & 0 \end{pmatrix} \quad (2.22)$$

and

$$G_0^\nu(\mathbf{q}, E) = \frac{1}{N^2} \sum_\mathbf{k} \frac{n_F^{(\nu)}(\mathbf{k+q}) + n_B^{(\nu)}(\mathbf{k})}{E - E_{\mathbf{k+q,k}}^{\nu\nu}} \quad (2.23)$$

is the bare Green's function resulting from the Hamiltonian (2.9) without interband hopping. Eventually, we find the RPA form of the single-electron Green's function in slave-boson mean-field theory,

$$G^{\nu\nu}(\mathbf{q}, E) = \frac{G_0^\nu(\mathbf{q}, E)}{1 - (g_\mathbf{q})^2 G_0^1(\mathbf{q}, E) G_0^2(\mathbf{q}, E)}. \quad (2.24)$$

## 2.3 Electronic Spectral Properties

To analyze the electronic spectrum, the total spectral density in momentum space associated with the Green's function (2.24),

$$S(\mathbf{q}, E) = -\frac{1}{\pi}\mathrm{Sp}\{\mathrm{Im}[G^{\nu\nu}(\mathbf{q}, E)]\}, \qquad (2.25)$$

is evaluated numerically for a hole concentration $x = 0.33$ and an inter-band hopping $t'/t = 0.15$ - the parameters fitting the dHvA Fermi surface [42]. For $\mathbf{q}$ along the $x$ or the $y$ axis in momentum space, $g_\mathbf{q}$ vanishes and the denominator of Eq. (2.24) is equal to 1 without any modifications to the original one-dimensional situation. In contrast, for $\mathbf{q}$ along the diagonal, $|g_\mathbf{q}|$ is maximal and the renormalization of the Green's function is strongest. Indeed, the recombination of spinons and holons into quasiparticles is most pronounced here, as is visible in the band hybridization and the evolution of quasiparticle weight as shown below.

First, we examine the spectrum along the main symmetry axis in the Brillouin zone (BZ) as denoted by the dashed line in Fig. 2.2(a) for the two temperatures $T = 0.1t$ and $T = 0.005t$, respectively (see Fig. 2.3). At high temperature, the spectrum resembles the two-particle continuum of a spinon and a holon, whose boundary is marked by the dotted lines in Fig. 2.3. At low temperature, the spectral weight shifts into two quasiparticle bands with a sharp momentum-energy relation. The resulting bands have the shape which is expected from a simple tight-binding model taking nearest- and next-nearest-neighbor hybridization into account, however, with a strongly enhanced mass ($m^*/m \approx 3$). Note that, since we ignore the lifetime effect due to the scattering among the resulting quasiparticles, these quasiparticle states remain well defined even away from the Fermi energy.

The shift of spectral weight to the quasiparticle bands is further illustrated in Fig. 2.4 which shows a cut of the spectral density at $\mathbf{q} = (\pi/2, \pi/2)$ for different temperatures and inter-band couplings. We see in this plot that the (incoherent) peak around $E = -0.3t$ without an inter-band coupling splits into two well-defined peaks of which the one at lower energies - below the spinon-holon continuum - remains almost unchanged even for higher temperatures while the one at higher energies is washed out.

Second, we further analyze the $q$ dependence as well as the temperature depen-

Dimensional Crossover in $Sr_2RuO_4$

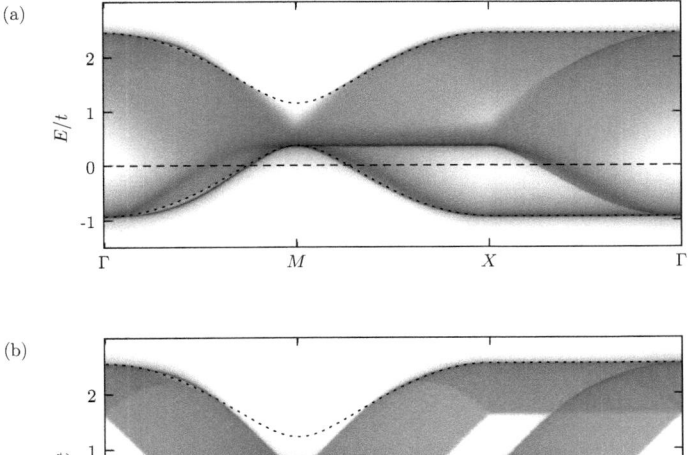

Figure 2.3: Density log-plot of the spectral-density function along the main symmetry axis for $T = 0.1t$ (a) and $T = 0.005t$ (b), respectively, showing the dispersion of the quasiparticles ('electrons'). The dotted lines mark the boundaries of the spinon-holon continuum and show that the emerging quasiparticle band is strongly renormalized. The dashed, horizontal line denotes the Fermi energy.

## 2.3 Electronic Spectral Properties

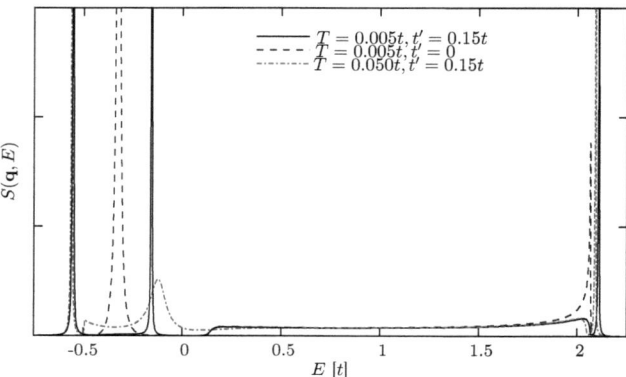

Figure 2.4: Comparison of the spectral density at $\mathbf{q} = (\pi/2, \pi/2)$ as a function of energy for the cases of low and high temperatures with inter-band hopping and low temperature without inter-band hopping.

dence of the quasiparticle spectral weight. Figure 2.5 shows density log-plots of the spectral function at the Fermi energy for different temperatures. At lowest temperature ($T = 0.001t$), the two-dimensional Fermi surface is well defined by quasiparticles [Fig. 2.5(a)] and agrees nicely with the ARPES picture in Fig. 2.1. With increasing temperature, the quasiparticle weight decreases and the Fermi surface becomes gradually faint. The retreat of the quasiparticle weight is clearly away from the [100] and [010] directions towards the [110] direction [Figs. 2.5.(b)-(d)]. A similar behavior of the quasiparticle weight was observed in ARPES experiments [28]. However, only the $\gamma$ band was studied in the [110] direction and thus, a complete comparison of the results is beyond the scope of our approach.

In our calculation, quasiparticles occur first along the diagonal momentum direction and extend slowly over a wider range of the emerging Fermi surface to form in an intermediate stage pieces of arc-like Fermi surfaces. To illustrate the evolution of quasiparticles at the Fermi surface, the angle dependence of the quasiparticle weight of the pocket around the $X$ point is shown for different temperatures in Fig. 2.6. Plotting the points of steepest descent on the Fermi

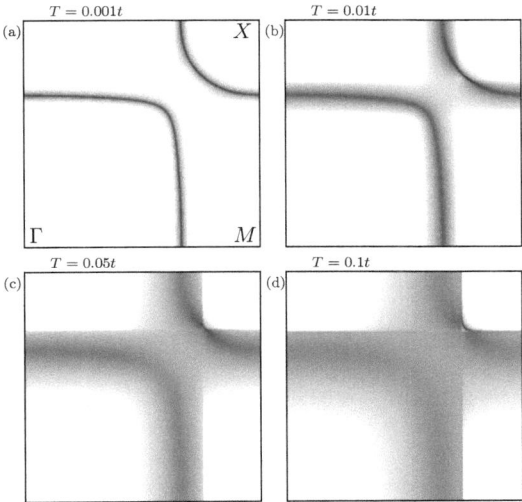

Figure 2.5: Density log-plots of the spectral density in momentum space at the Fermi energy for different temperatures. For $T = 0.001t$, a clear Fermi surface appears as expected for a hybridized, two-dimensional system. When increasing the temperature, the peaks of the spectral function get smeared and eventually, for $T = 0.1t$, we find a picture similar to that expected from a system of two decoupled one-dimensional bands.

## 2.4 c-Axis Transport

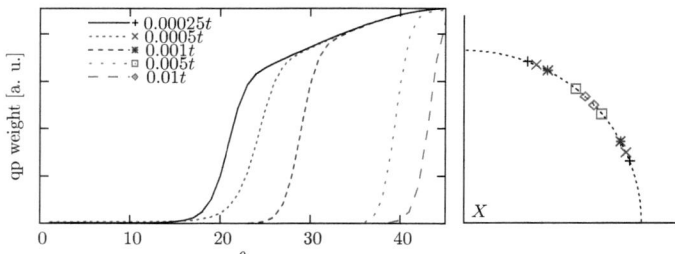

Figure 2.6: Quasiparticle weights as calculated from the spectral-density function in arbitrary units around the $X$ point as a function of the angle to the $x$ axis for different temperatures. When decreasing the temperature a growing arc of quasiparticle weight appears around the diagonals. To emphasize this, the figure on the right shows the endpoints of the arcs for different temperatures.

surface around the $X$ point, the right side of Fig. 2.6 nicely demonstrates the gradually increasing arc of quasiparticle weight. Within our calculation the Fermi arcs appear clearly only for the Fermi surface around the $X$ point ($\alpha$ Fermi surface). In contrast, for the Fermi surface centered around the $\Gamma$ point the rise of quasiparticle weight does not involve such a well defined arc feature.

Note that, unlike in other slave-boson discussions connected to superconductivity, the holon condensation is not involved in the formation of coherent quasiparticles here. The formation of the quasiparticles in the spectrum is caused by the inter-orbital hybridization which is taken into account in the RPA-like form of the Green's function (2.24). The picture obtained in our slave-boson mean-field approach should thus agree on a qualitative level with other descriptions of the dimensional crossover starting out from a Luttinger liquid in one dimension. The slave-boson mean-field theory carries the advantage of formal simplicity.

## 2.4 c-Axis Transport

In order to use a three-dimensional band description and calculate the c-axis resistivity in a Drude picture, the transport time of an electron moving in c direction should be sufficiently large as to go from layer to layer. Since the resistivity shows an extremely large anisotropy, this condition seems not to be met [43] and therefore, we analyze the temperature dependence of the charge transport along the $c$ axis within an interlayer-tunneling approach [44]. In this approach, the electric current is due to hopping of charge carriers from one layer to the next, where the layers act as reservoirs described by $\mathcal{H} = \mathcal{H}_0 + \mathcal{H}'$ from Eqs. (2.9) and (2.11) connected by the tunneling Hamiltonian

$$\mathcal{H}_T = \frac{1}{N^2} \sum_{\nu,\nu',s} \sum_{\mathbf{k},\mathbf{k}'} \sum_{\mathbf{p},\mathbf{p}'} \left( T^{\nu\nu'}_{\mathbf{kp}} \, f^{\dagger}_{\nu\mathbf{k}'+\mathbf{k}s} b_{\nu\mathbf{k}'} b^{\dagger}_{\nu'\mathbf{p}'} \, f_{\nu'\mathbf{p}'+\mathbf{p}s} + \text{h.c.} \right). \quad (2.26)$$

Operators with momentum $\mathbf{k}$ belong to the lower plane, while those with $\mathbf{p}$ belong to the upper plane and the tunneling matrix elements are

$$T^{\nu\nu'}_{\mathbf{kp}} = 4t''\delta_{\mathbf{kp}} \left\{ \delta_{\nu\nu'} \cos\frac{k_x + k_y}{2} + \sin\frac{k_x}{2} \sin\frac{k_y}{2} \right\} \quad (2.27)$$

for hopping from band $\nu$ to band $\nu'$ with $t'' \ll t$. This form follows from the analysis of the symmetry-allowed hopping paths between the layers possible via oxygen orbitals. In such a picture, a net current is obtained by shifting the chemical potential of one of the layers by an (infinitesimal) amount $eV$.

In general, we can express the total tunneling current in terms of a coherent and an incoherent tunneling current,

$$I_c^{\text{tot}} = I_c^{\text{coh}} + I_c^{\text{inc}}. \quad (2.28)$$

The coherent part consists of well-defined quasiparticles with life times long enough to tunnel from one layer to the next, while the incoherent part has to be expressed through independent spinon and holon hopping. Depending on the temperature range, we expect one or the other part to dominate the transport. For the maximum in the $c$-axis resistivity in Sr$_2$RuO$_4$ of around $T = 130\,\text{K}$, the crossover temperature corresponds to $T \approx 0.05t$ since $t \approx 0.3\,\text{eV}$. We therefore study a low-temperature region with $T \ll 0.05t$ and a high-temperature region with $T \gg 0.05t$.

### 2.4.1 Coherent transport

In the low-temperature limit, it was found above that the quasiparticle weight of the coherent spinon-holon pairs is growing with decreasing temperature and we

## 2.4 c-Axis Transport

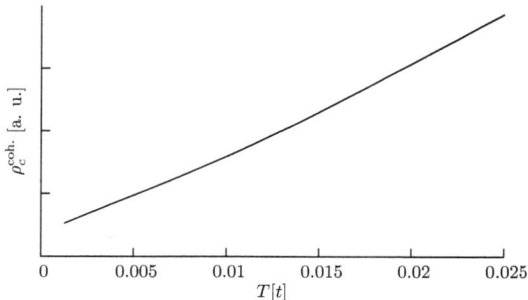

Figure 2.7: The resistivity in the low-temperature limit as calculated from Eq. (2.30). As expected, the decreasing quasiparticle weight in the low-temperature limit results in an increase in the resistivity as the temperature is increased.

expect the main contribution to the tunneling current to come from the coherent component. Following Cohen et al. [44], we determine the tunneling current by calculating the change in the charge of the lower layer, $N_\mathrm{L} = \sum_{\nu,\mathbf{k}} b_{\nu\mathbf{k}} b^\dagger_{\nu\mathbf{k}}$,

$$\dot{N}_\mathrm{L} = i[\mathcal{H}_\mathrm{T}, N_\mathrm{L}], \qquad (2.29)$$

and obtain

$$I = \frac{2e^2}{\hbar} V \sum_{\nu,\nu'} \sum_{\mathbf{k},\mathbf{p}} |T^{\nu\nu'}_{\mathbf{kp}}|^2 S^\nu_\mathrm{L}(\mathbf{k},0) S^{\nu'}_\mathrm{U}(\mathbf{p},0). \qquad (2.30)$$

Here, $S^\nu_{\mathrm{L(U)}}$ denotes the spectral density of the band $\nu$ in the lower (upper) layer. Using the spectral density as calculated from Eq. (2.25) this indeed yields an increasing resistivity due to the diminishing of quasiparticle weight upon growing temperature as is shown in Fig. 2.7. Note that, in this discussion, we ignore any other sources of temperature dependence, such as electron-electron and electron-phonon scattering.

### 2.4.2 Incoherent transport

In the high-temperature limit, Figs. 2.5 and 2.6 indicate that coherent quasiparticles which could tunnel between layers have disappeared. Therefore, the quasiparticle picture based on coherently paired spinons and holons as used for the discussion of the tunneling current above is inappropriate. We have to take a different approach whereby spin and charge degrees of freedom are deconfined. In the tunneling process of an electron, a spinon and a holon (charge carrier) have to be transferred simultaneously giving rise to incoherent transport [45]. We start with the Fermi Golden Rule to estimate the transfer probability of holons,

$$\Gamma^b_{i \to j} = \frac{2\pi}{\hbar} \sum_{\phi'} |\langle \phi_0 | \mathcal{H}_T | \phi' \rangle|^2 \delta(E_0 - E' + eV) \qquad (2.31)$$

with the many-body ground state $\phi_0$, $\phi'$ any excited state and $E_0$ and $E'$ the corresponding energies. Decoupling the holons and the spinons, we find for the total rate of holon tunneling

$$\Gamma^b_{i \to j} = 8 \frac{2\pi}{\hbar} |T_\perp|^2 \sum_{\mathbf{k},\mathbf{k'},\mathbf{q}} n^{(j)}_F(\mathbf{q})[1 - n^{(i)}_F(\mathbf{q})]$$
$$\times n^{(i)}_B(\mathbf{k'})[1 + n^{(j)}_B(\mathbf{k})] \delta(\omega_\mathbf{k} - \omega_{\mathbf{k'}} + eV) \qquad (2.32)$$

which requires that a spinon moves in the direction opposite to the holon. Here, $n^{(i)}_{F(B)}(\mathbf{k})$ denotes the Fermi (Bose) distribution with momentum $\mathbf{k}$ in the layer $i$ and $|T_\perp|^2$ is the (momentum independent) tunneling matrix element. The factor of 8 in the above formula is composed of a factor of 2 for spins and a factor of 4 because of the different bands involved. Since the distributions are the same on neighboring layers, we can write the net tunneling rate as

$$\Gamma^b_{\text{net}} = \frac{2\pi}{\hbar} 8 |T_\perp|^2 \int d\epsilon D_F(\epsilon) n_F(\epsilon)[1 - n_F(\epsilon)]$$
$$\times \int d\omega D_B(\omega) D_B(\omega + eV)[n_B(\omega + eV) - n_B(\omega)], \qquad (2.33)$$

where $D_{F(B)}$ is the density of states (DOS) of the spinons (holons). The integral over the Fermi distribution functions yields a factor of $D_F T$, with the DOS of the spinons $D_F$ taken to be constant at the spinon Fermi level. Taking the density of states of the holons $D_B$ to be constant, too, the integral over the Bose distributions can be written as

$$(D_B)^2 \int d\omega [n_B(\omega + eV) - n_B(\omega)] \approx D_B n_B^2 \kappa_B(T) eV, \qquad (2.34)$$

with the compressibility of the hard-core holons

$$\kappa_{\rm B}(T) = \frac{1}{n_{\rm B}^2} \frac{\partial n_{\rm B}(T,\mu)}{\partial \mu} \tag{2.35}$$

and the holon density $n_{\rm B}$.

Based on Eq. (2.33), the resulting current can now be estimated as

$$I = 8\frac{e^2}{\hbar}(2\pi)|T_\perp|^2 D_{\rm F} D_{\rm B} T n_{\rm B}^2 \kappa_{\rm B}(T) V, \tag{2.36}$$

leading to the conductivity

$$\sigma = 8\frac{e^2}{\hbar}(2\pi)|T_\perp|^2 D_{\rm F} D_{\rm B} T n_{\rm B}^2 \kappa_{\rm B}(T) \frac{c}{ab}. \tag{2.37}$$

The compressibility of the holons $\kappa_{\rm B}(T)$ corresponds to that of spinless fermions and is constant for temperatures much smaller than the characteristic energy of holons which is of the order $t\chi^f$. Thus, the temperature dependence of $\sigma$ in the range of interest is dominated by the term $D_{\rm F}T$ due to the spinon phase-space contribution which is increasing with growing temperature. Physically, this means that, since spinons have to be available for holons to move from one layer to another ("spinons act as a bookkeeping system"), an increased phase space for elastic spinon scattering leads to an increase in the conductance rather than the resistance. Therefore, we find a temperature dependence of an insulator, i.e., a decreasing resistivity, in the temperature range above $T = 0.05t$ as is shown in Fig. 2.8.

## 2.5 Conclusion

To understand the transport properties of $Sr_2RuO_4$ and the different regimes found in experiment, it is crucial to understand the underlying electronic band structure of the basal plane. The in-plane transport is dominated by the two-dimensional $\gamma$ band and thus, shows Fermi-liquid behavior up to room temperature. In contrast, for the $c$-axis transport, only the $\alpha$ and $\beta$ bands, stemming from the $d_{yz}$ and $d_{zx}$ Ru orbitals and thus having a weak interlayer hybridization, are relevant. The regime change of the $c$-axis transport in $Sr_2RuO_4$ can therefore

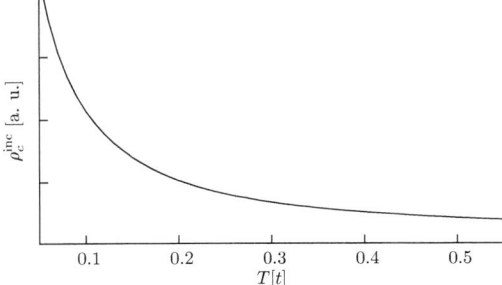

Figure 2.8: The resistivity in the high-temperature limit as calculated from Eq. (2.37). The possibility of elastically scatter on spinons dominates the holon transport resulting in a decreasing resistivity.

be interpreted in terms of a dimensional crossover (from one to two dimensions) for these bands in the basal plane. The discussion of the $c$-axis resistivity in this chapter shows that the slave-boson approximation can give a qualitative understanding and provides interesting insight into the change of transport properties. In particular, our results lead to a picture, where, with decreasing temperature, a Fermi surface appears through gradually extending arcs of finite quasiparticle weight. This kind of Fermi-surface evolution could be accessible to more specific ARPES experiments. In view of the progressive emergence of coherent quasiparticles the difference between the metallic and insulating behavior can be understood as a change between dominant coherent to dominant incoherent tunneling as temperature is raised.

## 2.5 Conclusion

# Chapter 3

# Metamagnetism and Nematicity in $Sr_3Ru_2O_7$

> *This chapter aims at explaining the magnetic-field-direction anisotropy of the response of $Sr_3Ru_2O_7$. Based on a microscopic tight-binding model, a metamagnetic transition is introduced by means of a van Hove singularity scenario. Taking the rotations of the O octahedra into account, an additional term in the Hamiltonian is derived microscopically, a staggered spin-orbit coupling. This term leads to an anisotropic response to a magnetic field with in-plane fields leading to a (commensurate) spin-density wave and the expected anisotropic appearance of the intermediate phase.*

## 3.1 Introduction

This chapter deals with the properties of $Sr_3Ru_2O_7$ in a magnetic field and the anisotropy in the response to it. However, before discussing the effects of an applied field, we first review some zero-field low-temperature properties. Under such conditions, $Sr_3Ru_2O_7$ is a paramagnetic metal with strongly two-dimensional character which manifests itself in a resistivity anisotropy of $\rho_c/\rho_{ab} \sim 300$ at $0.3\,\text{K}$ [19]. For low enough temperatures, the resistivity shows a Fermi-liquid-like $T^2$ dependence. The relatively high specific heat of $\gamma = 110\,\text{mJ}/(\text{K}^2\,\text{mol}$

## 3.1 Introduction

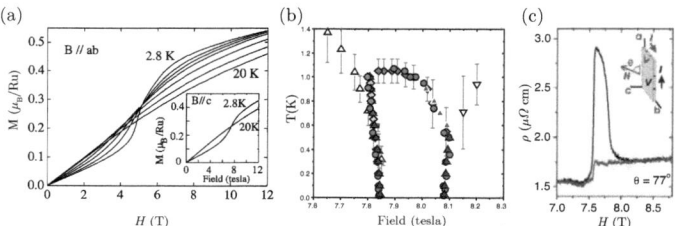

Figure 3.1: Experimental data on $Sr_3Ru_2O_7$ leading to the characterization of the phase diagram. (a) Metamagnetic transition for fields $H \parallel ab$ and $H \perp ab$ (inset). (b) Boundary of the intermediate phase as obtained from several thermodynamic measurements. (c) Anisotropic in-plane resistivity within the nematic phase for a field slightly tilted away from the $c$ axis (cf. inset). For details on the individual graphs, see references [2, 25, 47].

Ru) as compared to e.g., $\gamma = 38\,\mathrm{mJ/(K^2\,mol\,Ru)}$ for $Sr_2RuO_4$ [20] is a sign of a high density of states at the Fermi energy. In addition, a pronounced peak in the zero-field susceptibility as a function of temperature around $T_{max} = 16\,\mathrm{K}$ indicates a peak in the DOS close to the Fermi energy. Regarding its tendency to order, multiple experiments indicate a proximity of $Sr_3Ru_2O_7$ to ferromagnetism. A Wilson ratio of 10, an induced ferromagnetic instability under uniaxial pressure [19], and also inelastic neutron-scattering measurements [46] are pointing in this direction. This is also supported by early band-structure calculations [21].

As was first discussed by Wohlfarth and Rhodes [48], such a situation can result in metamagnetic behavior, a superlinear rise in the magnetization over a narrow region of applied magnetic field $H$. This phenomenon was observed in a number of systems [49] and was also found in $Sr_3Ru_2O_7$ [47] with a critical field $H_c$, however, that depends strongly on the angle of the field versus the $ab$ plane. While for $\theta = 0°$ the critical field $H_c \sim 5.1\,\mathrm{T}$, it rises to $H_c \sim 8\,\mathrm{T}$ for $\theta = 90°$ [see Fig. 3.1(a)]. Measurements of the differential magnetic susceptibility of high-quality single crystals grown using a floating zone technique ($\rho_{res} = 2.4\,\mu\Omega\,\mathrm{cm}$) in addition showed that the occurring transition is first order for in-plane fields and $T < T^* \approx 1.25\,\mathrm{K}$, while it is only a crossover for fields parallel to $c$ [50]. It was therefore suggested that the field angle could be used as a tuning parameter for a line of first-order transitions that goes to $T^* = 0\,\mathrm{K}$ around $\theta = 80°$, thus, realizing a QCEP [23] [see schematic phase diagram in Fig. 3.2(a)].

With a further improvement of crystal quality and associated residual resistivities down to $\rho_{\mathrm{res}} < 1\,\mu\Omega\,\mathrm{cm}$, Grigera et al. performed a new set of experiments attempting to describe the putative quantum critical endpoint. However, before reaching the QCEP, the system entered a new phase first characterized by a strongly enhanced resistivity and bound by two consecutive jumps in the magnetization. Measurements of several thermodynamic properties allowed to exactly determine the boundaries of this novel phase as shown in Fig. 3.1(b) [25, 51]. A detailed analysis of the resistivity of this intermediate phase revealed a large in-plane transport anisotropy with a $C_2$ symmetry [Fig. 3.1(c)] [2]. As there was no detectable structural change of the underlying lattice, it was argued that this was due to an induced anisotropic electronic state with a symmetry-breaking Fermi surface deformation similar to a Pomeranchuk instability [52]. That this kind of a phase, also called electronic nematic phase in analogy to liquid crystal phases, can lead to two consecutive metamagnetic transitions had already been demonstrated in a paper by Kee and Kim [53]. In addition, the anomalous $T$ dependence of the susceptibility $\chi$ and the specific heat coefficient $\gamma$ could be explained in such a scenario [54]. Moreover, the two-fold degeneracy of the nematic phase allows for domain formation, such that domain-wall scattering could account for the increased resistivity of the intermediate phase [55].

As in the case of $Sr_2RuO_4$ in the previous chapter, the electronic structure of a single layer of $Sr_3Ru_2O_7$ is dominated by the bands originating from the $4d$ $t_{2g}$ orbitals $d_{yz}$, $d_{zx}$ and $d_{xy}$ hybridizing with the O $2p$ orbitals. In a simple approximation this leads to the two quasi-one-dimensional bands with mainly $d_{yz}$ and $d_{zx}$ character and a two-dimensional band stemming from the $d_{xy}$ orbital. Due to the interlayer coupling these three bands are then additionally split resulting in 6 bands. An important consequence of the bilayer splitting is that one of the two bands coming from the $d_{xy}$ orbitals is shifted closer to the van Hove singularity (vHS). This was also confirmed by recent ARPES measurements [56]. A chemical potential in the vicinity of a vHS is the condition for the scenario described by Binz and Sigrist for a metamagnetic transition [57]. Proximity to a vHS could also lead to a nematic phase accompanying a metamagnetic transition as described by Grigera et al. [25]. The anisotropy in the critical field strength

## 3.1 Introduction

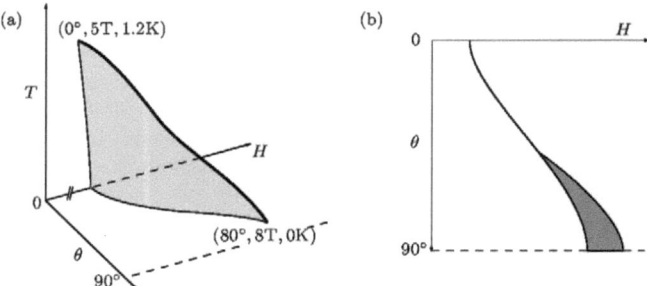

Figure 3.2: (a) Schematic phase diagram of $Sr_3Ru_2O_7$ without the intermediate phase for fields $H$ applied with an angle $\theta$ versus the $ab$ plane. The surface represents first-order transitions separating a region with low (spin) polarization from a region with high polarization. The thick black line connecting $(\theta, H, T) = (0°, 5\,T, 1.2\,K)$ and $(80°, 8\,T, 0\,K)$ is a line of critical endpoints. (b) Schematic $H$-$\theta$ phase diagram at $T = 0$ showing the nematic phase (gray area) for fields applied close to the $c$-axis direction. For details see main text.

could, in such a picture, be explained by spin-orbit coupling effects similar to [58] leading to an anisotropic effective $g$ factor.

A different route is taken by Raghu et al. [59] and Lee et al. [60] who studied a model where the metamagnetic transition comes from the two (bilayer-split) one-dimensional bands. The anisotropy is then again introduced by considering spin-orbit interaction on the Ru sites and the nematic phase can be understood as an orbital ordering among these one-dimensional bands. Both these routes suffer, however, from a shortcoming: even though they can describe the existence of a nematic phase and a dependence of $H_c$ on the field angle, they cannot explain why the nematic phase occurs only for fields almost parallel to the crystalline $c$ axis [see schematic phase diagram in Fig. 3.2(b)].

This point is addressed in the present work. We study a model based on a two-dimensional band in a single layer originating from the $d_{xy}$ orbitals. The bilayer effects are only taken into account by placing this band closer to the vHS. Starting from this, we consider the effect of a lattice distortion in the planes. As was first shown by Shaked and coworkers in a neutron-scattering experiment [61], $Sr_3Ru_2O_7$ does not realize the high symmetry structure shown in Fig. 1.1, but the O octahedra are rotated by 6.8° due to the ionic size of $Sr^{2+}$. The rotation of the octahedra is around the $c$ axis and is in opposite direction for corner-sharing octahedra. The symmetry of the crystal is thus reduced and the unit cell changes from body-centered tetragonal to a $\sqrt{2}\times\sqrt{2}$ larger face-centered orthogonal cell with $a$ and $b$ almost identical, however [62]. Such a structure was recently also studied by Puetter $et.$ $al.$ [63], but the doubling of the unit cell was only introduced by adding an effective lattice potential. Starting from a microscopic Hamiltonian and considering relativistic corrections, we show how the rotated oxygen octahedra introduce a staggered spin-orbit coupling, an effect similar to the Dzyaloshinskii-Moriya interaction for localized spins [64, 65], here, however, for itinerant electrons. For magnetic fields applied in the plane, this adds a component with wave vector $\mathbf{Q}=(\pi,\pi)$ to the static susceptibility. The induced commensurate spin-density wave (SDW) opens gaps in the Fermi surface close to the van Hove points which has an impact on the occurrence of any instability that emerges due to the proximity to a vHS. Here, we choose to describe the low-temperature phase as an electronic nematic state to examine this aspect, since it relies on the presence of the vHS and represents one of the most promising candidates for the intermediate phase. In order to discuss the essential influence of the staggered SOC on the phase diagram we adopt a mean-field approach, with the shortcoming that critical fluctuations are not included well. While, in particular, quantum critical fluctuations represent an intriguing part of the phenomenology of this metamagnetic transition, we assume that they are not essential to understand the basic effects due to SOC and lie beyond the scope of this study. Note that we do not include the spin-orbit coupling on the Ru ions. This type of SOC couples different bands and requires the incorporation of all three $t_{2g}$ bands. As mentioned above, the main effect of such a term would be an anisotropic $g$ tensor and could not explain the anisotropy in the appearance of the intermediate phase.

This chapter is organized as follows: in the next section, we introduce our model based on a three-band Hamiltonian consisting of the Ru $4d_{xy}$ orbital and the in-plane O $2p$ orbitals. After reducing this model to an effective one-band model, we analyze the effect of a rotation of the oxygen octahedra on it. On-site interactions are then treated within mean-field theory and an additional applied magnetic field is considered. The resulting model is then studied in a next section. Following Metzner et al. [66], we add a forward-scattering term in section 3.4 to allow for a nematic phase and study the influence of the staggered SOC to this phase. In a last section, we discuss and summarize our findings.

## 3.2 Microscopic Derivation of the Hamiltonian

### 3.2.1 Basic hopping Hamiltonian

The model system we want to analyze here is based on a single $RuO_4$ layer. Instead of introducing a second layer and thus having each band split, we take the bilayer splitting into account by simply shifting the band closer to the vHS. In contrast to the previous chapter, where we started from a Hamiltonian with the oxygen degrees of freedom already integrated out, thus only taking Ru $d$ orbitals into account, the starting point in this chapter is a three-band tight-binding model including the in-plane $4d$ Ru orbital ($d_{xy}$) as well as the two in-plane $2p$ O orbitals ($p_x$ and $p_y$). These orbitals have on-site energies $E_d$ and $E_p = E_d - \Delta$, respectively. This underlying model is necessary to adequately incorporate the symmetry of this system. In this model, an electron can hop from a $d_{xy}$ orbital in $x$ ($y$) direction to a $p_y$ ($p_x$) oxygen orbital with the hopping integral $t_{dp_y}$ ($t_{dp_x}$) and vice versa. Additionally, due to strong hybridization of the oxygen $2p$ orbitals, electrons can hop between neighboring oxygen orbitals. This leads to a Hamiltonian of the general form

$$\mathcal{H}_{3b} = \sum_s \vec{C}_s^\dagger \begin{pmatrix} E_d & \tilde{t}_{dp_x} & \tilde{t}_{dp_y} \\ \tilde{t}_{dp_x} & E_{p_x} & \tilde{t}_{pp} \\ \tilde{t}_{dp_y} & \tilde{t}_{pp} & E_{p_x} \end{pmatrix} \vec{C}_s, \tag{3.1}$$

where $\vec{C}^\dagger = (d^\dagger, p_x^\dagger, p_y^\dagger)$ are the creation operators for the above mentioned orbitals. Care has to be taken of the different signs of the hopping integrals due to the relative phases of the orbital wave functions indicated by the tildes [see Fig. 3.3(a)].

To reduce our model to one band, we construct an effective Hamiltonian only living at the Ru sites by integrating out the high-energy degrees of freedom [67,

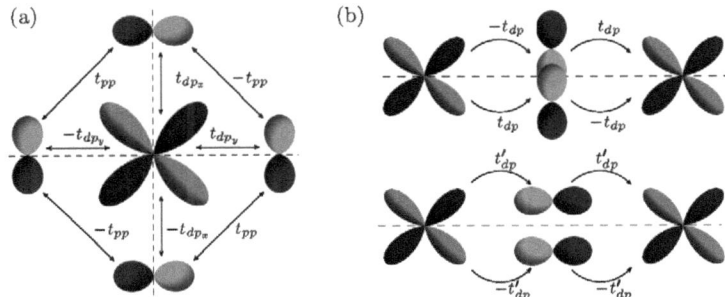

Figure 3.3: (a) Different hoppings in the three-band model of Ru $4d_{xy}$ and O $2p_x$ / $p_y$ orbitals. The relative signs of the hopping integrals come from the relative phases of the orbital wave-functions. (b) Possible hoppings for the three-band model in the case of rotated O octahedra. The sign change in the hopping integrals is again due to the relative phases of the Wannier functions and holds in first order.

68],

$$\mathcal{H}_{\text{eff}} = \sum_p \frac{\mathcal{H}_{3b}|p\rangle\langle p|\mathcal{H}_{3b}}{E_d - E_p} + \sum_{p,p'} \frac{\mathcal{H}_{3b}|p\rangle\langle p|\mathcal{H}_{3b}|p'\rangle\langle p'|\mathcal{H}_{3b}}{(E_d - E_p)(E_d - E_{p'})}, \quad (3.2)$$

where the sums run over all oxygen orbitals on all sites. This leads to a simple hopping Hamiltonian

$$\mathcal{H}_0 = -t \sum_{\langle i,j \rangle} \sum_s c^\dagger_{is} c_{js} - t' \sum_{(i,j)} \sum_s c^\dagger_{is} c_{js}, \quad (3.3)$$

where $c^\dagger_{is}$ creates an electron at Ru site $i$ with spin $s$, $\langle i,j \rangle$ denotes nearest neighbors and $(i,j)$ next-to-nearest neighbors. To lowest order, the hopping integrals in this effective Hamiltonian then read

$$t = \frac{t_{dp}^2}{\Delta}, \quad (3.4)$$

$$t' = \frac{t_{dp}^2 t_{pp}}{\Delta^2} \quad (3.5)$$

## 3.2 Microscopic Derivation of the Hamiltonian

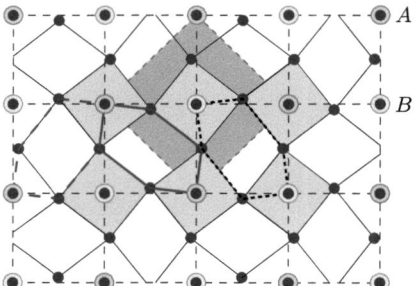

Figure 3.4: The RuO$_4$ plane with rotated oxygen octahedra leading to a doubling of the unit cell as indicated by the large dashed square. To emphasize that the rotation is only around the $c$ axis, the apical oxygens are also shown on top of the ruthenium sites. The two distinct lattice sites are denoted as $A$ (darker gray) and $B$ (lighter gray). The inversion symmetry of the bonds between Ru ions is broken leading to a staggered spin-dependent nearest-neighbor hopping. The next-nearest-neighbor hopping is still isotropic and the same for both lattice sites as is depicted by the different equivalent hopping-paths (bold solid, dashed lines for $A$ to $A$ and dotted line for $B$ to $B$).

with $t_{dp_x} = t_{dp_y} = t_{dp}$.

In the following, we additionally take the rotation of the oxygen octahedra into account. This rotation shifts the oxygen in between neighboring Ru sites into an off-center position, breaking the inversion symmetry on the bonds. Since this shifting is in a staggered fashion, we obtain a bipartite lattice leading to a $\sqrt{2} \times \sqrt{2}$ larger unit cell separating $A$ and $B$ sublattices (see Fig. 3.4). In the simple tight-binding model we have considered so far, this aspect is not reflected, yet. Before deriving a symmetry-reducing term for the Hamiltonian from microscopics, we first shortly discuss the situation from a more phenomenological point of view. To illustrate the special form to expect for this term, we consider the O ion as yielding an electric field transverse to the direct path of the electrons moving from one Ru site to the next. Introducing relativistic corrections, we obtain a spin-orbit coupling of the form

$$\mathcal{H}^{\text{soc}} = \tilde{\alpha}(\vec{k} \times \vec{n}) \cdot \vec{\sigma}, \tag{3.6}$$

where $\vec{n}$ is the in-plane normal vector to the electron motion and the sign of $\tilde{\alpha}$ is site dependent. Therefore, we expect to find a (staggered) SOC of Rashba type.

From a microscopic point of view, a consequence of the shifted O ions is that the formerly symmetry-forbidden hopping in $x$ ($y$) direction via $p_x$ ($p_y$) orbitals is now possible with matrix element $t'_{dp}$ as is depicted in Fig. 3.3(b). Thus, we might include spin-orbit coupling at the oxygen site, $\mathcal{H}^{O-2p} = \lambda L_z S_z$, which mixes the $p_x$ and $p_y$ orbitals. On the oxygen sites, we therefore need to change to eigenfunctions of the SOC $|\pm\rangle$, with $\mathcal{H}^{O-2p}|\pm\rangle = \pm\lambda s|\pm\rangle$ ($s = \pm 1$ is the spin index).

As an example, we write the total Hamiltonian for the sublattice $A$ for the $x$ direction,

$$\begin{aligned}
\mathcal{H}_A^{(x)} = &-\sum_{j\in A}\sum_s (\tilde{t} d_{js}^\dagger p_{+,j+\hat{x}/2s} + \tilde{t}^* d_{js}^\dagger p_{+,j-\hat{x}/2s} + \text{h.c.}) \\
&-\sum_{j\in A}\sum_s (\tilde{t}^* d_{js}^\dagger p_{-,j+\hat{x}/2s} + \tilde{t} d_{js}^\dagger p_{-,j-\hat{x}/2s} + \text{h.c.}) \\
&-\sum_{\nu=\pm}\sum_{as}(\Delta \pm \lambda s) p_{\nu,as}^\dagger p_{\nu,as},
\end{aligned} \tag{3.7}$$

where $p_{\pm,j+\hat{x}/2s}^\dagger$ creates an electron at the oxygen site $j+\hat{x}/2$ in the state $|\pm\rangle$ with spin $s$ and

$$\tilde{t} = \frac{t'_{dp} - i t_{dp}}{\sqrt{2}}. \tag{3.8}$$

Applying perturbation theory in the from of Eq. (3.2) to this Hamiltonian to construct an effective model again, we find

$$\begin{aligned}
\langle A|\mathcal{H}_{\text{eff}}|B\rangle &= \frac{\tilde{t}^2}{\Delta+\lambda s} + \frac{(\tilde{t}^*)^2}{\Delta-\lambda s} \\
&= -(t_{pd}^2 - t_{pd}'^2)\frac{\Delta}{\Delta^2-\lambda^2} + is\frac{2\lambda t_{pd} t'_{pd}}{\Delta^2-\lambda^2} \\
&= -t + i\alpha s
\end{aligned} \tag{3.9}$$

## 3.2 Microscopic Derivation of the Hamiltonian

for the hopping integral from a site $A$ to a site $B$ in the positive $x$ direction. Hence, we have a total Hamiltonian $\mathcal{H}^{\text{hop}} = \mathcal{H}_0 + \mathcal{H}^{\text{soc}}$ with a nn-and nnn-hopping Hamiltonian $\mathcal{H}_0$ and a staggered spin-dependent hopping in the form of a staggered SOC of Rashba type,

$$\mathcal{H}^{\text{soc}} = \sum_{ss'} \Big[ -i\alpha \sum_{j\in A} \sum_{\hat{a}=\hat{x},\hat{y}} (c^\dagger_{j+\hat{a}s} c_{js'} - c^\dagger_{js'} c_{j+\hat{a}s}) \sigma^z_{ss'} \quad (3.10)$$

$$+ i\alpha \sum_{j\in B} \sum_{\hat{a}=\hat{x},\hat{y}} (c^\dagger_{j+\hat{a}s} c_{js'} - c^\dagger_{js'} c_{j+\hat{a}s}) \sigma^z_{ss'} \Big]. \quad (3.11)$$

Note that the nnn-hopping integrals, even though renormalized, do not become anisotropic, which can be deduced from geometrical considerations as indicated in Fig. 3.4.

The bipartite lattice introduces a wave-vector $\mathbf{Q} = (\pi, \pi)$ with which the total Hamiltonian in momentum space reads

$$\mathcal{H}^{\text{hop}} = \sum_{ss'} \sum_{\mathbf{k}}{}' \vec{c}^\dagger_{\mathbf{k}s} \begin{pmatrix} \varepsilon^{(1)}_{\mathbf{k}} + \varepsilon^{(2)}_{\mathbf{k}} - \mu & ig_{\mathbf{k}}\sigma^z_{ss'} \\ -ig_{\mathbf{k}}\sigma^z_{ss'} & -\varepsilon^{(1)}_{\mathbf{k}} + \varepsilon^{(2)}_{\mathbf{k}} - \mu \end{pmatrix} \vec{c}_{\mathbf{k}s'}, \quad (3.12)$$

where $\vec{c}^\dagger_{\mathbf{k}s} = (c^\dagger_{\mathbf{k}s}, c^\dagger_{\mathbf{k}+\mathbf{Q}s})$. Here,

$$\varepsilon_{\mathbf{k}} = -2t(\cos k_x + \cos k_y) - 4t' \cos k_x \cos k_y = \varepsilon^{(1)}_{\mathbf{k}} + \varepsilon^{(2)}_{\mathbf{k}} \quad (3.13)$$

are the hopping energies for nn- and nnn-hopping and $g_{\mathbf{k}} = 2\alpha(\cos k_x + \cos k_y)$ is the form factor for the spin-orbit coupling. The prime in the $\mathbf{k}$ summation indicates that the summation runs only over the reduced first BZ. As this restriction holds for all subsequent $\mathbf{k}$ sums we omit the prime in the following.

From Eq. (3.12), we see that the staggered SOC hybridizes states with $\mathbf{k}$ and $\mathbf{k}+\mathbf{Q}$. It is now convenient to introduce Pauli matrices $(\tau^0, \vec{\tau})$ in momentum space $\{\mathbf{k}, \mathbf{k}+\mathbf{Q}\}$ such that we can write the Hamiltonian as

$$\mathcal{H}^{\text{hop}} = \sum_{\mathbf{k}} \sum_{s,s'} \vec{c}^\dagger_{\mathbf{k}s} \mathcal{H}^{\text{hop}}_{\mathbf{k}ss'} \vec{c}_{\mathbf{k}s'} \quad (3.14)$$

with

$$\mathcal{H}^{\text{hop}}_{\mathbf{k}ss'} = (\varepsilon^{(2)}_{\mathbf{k}} - \mu)\sigma^0_{ss'}\tau^0 + \varepsilon^{(1)}_{\mathbf{k}}\sigma^0_{ss'}\tau^3 - g_{\mathbf{k}}\sigma^z_{ss'}\tau^2. \quad (3.15)$$

Diagonalizing this Hamiltonian yields two (spin-degenerate) bands,

$$\xi_{\alpha \mathbf{k} s} = \epsilon_{\mathbf{k}}^{(2)} + (-1)^\alpha \sqrt{(\epsilon_{\mathbf{k}}^{(1)})^2 + g_{\mathbf{k}}^2}, \tag{3.16}$$

where $\alpha = 1, 2$. As can be seen in Fig. 3.11(a), where the Fermi surface is plotted, the first BZ is folded back. This is in accordance with the doubling of the unit cell introduced by the rotated oxygen octahedra.

### 3.2.2 Magnetization and on-site interaction

Before adding an on-site interaction to the Hamiltonian (3.15), we first want to examine the effect of an applied magnetic field. This is necessary for the later dealing with the interaction within a mean-field approach, as we need to choose an adequate spin-quantization axis. From the form of Eq. (3.15) we first see that a magnetic field in $z$ direction is a mere spin-dependent shift of the chemical potential. The response of the system is thus a simple polarization. However, for in-plane fields, the staggered spin-orbit coupling introduces a coupling of homogeneous magnetic fields to a staggered magnetization, i.e., a commensurate SDW. This means that the static spin-susceptibility has a component with wave-vector $\mathbf{Q}$. To see this explicitly, we add a Zeeman term of the form

$$\mathcal{H}^Z = -g\mu_B [\vec{H}_\mathbf{0} \cdot \vec{S}(\mathbf{0}) + \vec{H}_\mathbf{Q} \cdot \vec{S}(\mathbf{Q})] \tag{3.17}$$

with the spin operators

$$\vec{S}(\mathbf{q}) = \frac{1}{2} \sum_{\mathbf{k}} c^\dagger_{\mathbf{k}+\mathbf{q} s} \vec{\sigma}_{ss'} c_{\mathbf{k} s'} = \sum_{\mathbf{k}} \vec{S}_\mathbf{k}(\mathbf{q}), \tag{3.18}$$

the Bohr magneton $\mu_B$ and the Landé factor $g$. This corresponds to a homogeneous and a staggered magnetic field, in accordance with the structure of the Hamiltonian given in Eq. (3.15), and can, thus, be written as

$$\mathcal{H}^Z_{\mathbf{k} ss'} = (\vec{h}_\mathbf{0} \cdot \vec{\sigma}\tau^0 + \vec{h}_\mathbf{Q} \cdot \vec{\sigma}\tau^1). \tag{3.19}$$

Here, we have introduced $\vec{h}_{\mathbf{0}/\mathbf{Q}} = \vec{H}_{\mathbf{0}/\mathbf{Q}}/H_0$, where $H_0 = -2 \cdot 10^{-4} t/(g\mu_B)$.

It is now straightforward to calculate the magnetic response of the system to an applied field by using the thermodynamic relation

$$\langle m^i_{\mathbf{0}/\mathbf{Q}} \rangle = -\frac{\partial}{\partial h^i_{\mathbf{0}/\mathbf{Q}}} F(T, \vec{h}_\mathbf{0}, \vec{h}_\mathbf{Q}, n), \tag{3.20}$$

## 3.2 Microscopic Derivation of the Hamiltonian

where $F(T, \vec{h}_\mathbf{0}, \vec{h}_\mathbf{Q}, n)$ is the free energy, $\langle m_\mathbf{0}^i \rangle$ is the homogeneous magnetization pointing in the $i$ direction and $\langle m_\mathbf{Q}^i \rangle$ corresponds to a staggered magnetization. Adding the Zeeman term (3.19) to the Hamiltonian (3.15), we find a finite staggered magnetization in $y$ direction for the case of a homogeneous field applied in $x$ direction,

$$\langle m_\mathbf{Q}^y \rangle = \frac{1}{2} \sum_{\alpha=1,2} \sum_{\beta=\pm} \sum_{\mathbf{k}} n_\mathrm{F}(\xi_{\alpha\beta,\mathbf{k}}) \frac{(-1)^\alpha g_\mathbf{k}}{\sqrt{(h_0^x \pm \epsilon_\mathbf{k}^{(1)})^2 + g_\mathbf{k}^2}}. \tag{3.21}$$

In the above equation,

$$\xi_{\alpha\pm,\mathbf{k}} = \epsilon_\mathbf{k}^{(2)} + (-1)^\alpha \sqrt{(h_0^x \pm \epsilon_\mathbf{k}^{(1)})^2 + g_\mathbf{k}^2} \tag{3.22}$$

are the four energy bands in a homogeneous field in $x$ direction and $n_\mathrm{F}(\xi)$ is the Fermi distribution.

We are now in the position to treat an on-site interaction term within mean-field theory. If we introduce such a term in the Hamiltonian,

$$\mathcal{H}^U = U \sum_i n_{i\uparrow}^{\vec{a}} n_{i\downarrow}^{\vec{a}}, \tag{3.23}$$

we first have to choose an appropriate spin quantization axis (indicated by the superscript $\vec{a}$). For simplicity, we only consider the two cases of an applied magnetic field in $z$ and in $x$ direction in the following.

### Field applied in $z$ direction

Since a field perpendicular to the plane does not couple to any staggered magnetization, the first case is the straightforward text-book case. The quantization axis is the $z$ axis and we write

$$\mathcal{H}^U = U \sum_i \left[ \frac{(n_{i\uparrow} + n_{i\downarrow})^2}{4} - \frac{(n_{i\uparrow} - n_{i\downarrow})^2}{4} \right]. \tag{3.24}$$

Since we do not expect large fluctuations in the charge density, the first term is a constant and thus, the interaction can be written as

$$\mathcal{H}^U = -U \sum_i S_i^z S_i^z \tag{3.25}$$

with the spin polarization $S_i^z = (n_{i\uparrow} - n_{i\downarrow})/2$. Applying mean-field theory to this expression and changing to momentum space yields

$$\mathcal{H}^U = -2UM^z \sum_{\mathbf{k}} S_{\mathbf{k}}^z(\mathbf{0}) + UN(M^z)^2 \qquad (3.26)$$

with $M^z = \langle S_i^z \rangle$ independent of site $i$. Therefore, this leads to the additional term

$$\mathcal{H}_{\mathbf{k}}^U = -UM^z \sigma^z \tau^0 + U(M^z)^2, \qquad (3.27)$$

with which the total Hamiltonian now reads

$$\mathcal{H}_{\mathbf{k}ss'} = \mathcal{H}_{\mathbf{k}ss'}^{\text{hop}} + \tilde{h}_0^z \sigma^z \tau^0. \qquad (3.28)$$

The hopping Hamiltonian is given in Eq. (3.15) and we defined an effective magnetic field $\tilde{h}_0^z = h_0^z - UM^z$. This Hamiltonian has four bands,

$$\xi_{\alpha \mathbf{k} s} = \epsilon_{\mathbf{k}}^{(2)} + s\tilde{h}_0^z + (-1)^\alpha \sqrt{g_{\mathbf{k}}^2 + (\epsilon_{\mathbf{k}}^{(1)})^2} - \mu \qquad (3.29)$$

and, via the grand-canonical potential per lattice site,

$$\omega = -T \sum_{\alpha=1,2} \sum_{\mathbf{k},s} \log[1 + \exp(-\xi_{\alpha \mathbf{k} s}/T)] + U(M^z)^2, \qquad (3.30)$$

the self-consistency equations can be derived in the form

$$n = \frac{1}{N} \sum_{\alpha=1,2} \sum_{\mathbf{k},s} n_\text{F}(\xi_{\alpha \mathbf{k} s}), \qquad (3.31)$$

$$M^z = \frac{1}{2N} \sum_{\alpha=1,2} \sum_{\mathbf{k},s} s\, n_\text{F}(\xi_{\alpha \mathbf{k} s}). \qquad (3.32)$$

**Field applied in $x$ direction**

The case of a field in $x$ direction is slightly more involved since we expect an additional response in the form of a staggered magnetization in $y$ direction leading to a canted magnetization in $x$ direction. Therefore, we define an angle $\phi$ for the canting angle and denote with $M$ the total moment (see Fig. 3.5). In Eq. (3.23), the quantization axis $\hat{a}$ and $\hat{b}$ for the spins on lattice sites $A$ and $B$, respectively,

## 3.2 Microscopic Derivation of the Hamiltonian

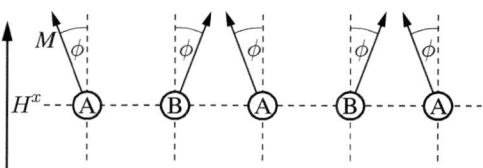

Figure 3.5: Real space schematic of the magnetic order for the case of an applied field in the $xy$ plane: Due to the staggered spin-orbit coupling, the magnetization is also staggered with respect to the sublattice sites $A$ and $B$ with order parameter $M$ and $\phi$ for the total moment and canting angle, respectively.

should therefore be perpendicular to $z$ and tilted away from the $x$ direction by the angle $\phi$. Decoupling the charge- and spin-density parts of Eq. (3.23) and keeping again only the latter, we find

$$\begin{aligned}
\mathcal{H}^U &= -U \sum_{i \in A} S_i^{\hat{a}} S_i^{\hat{a}} - U \sum_{i \in B} S_i^{\hat{b}} S_i^{\hat{b}} \\
&= UNM^2 - 2UM \sum_{i \in A} (S_i^x \cos\phi + S_i^y \sin\phi) \\
&\quad - 2UM \sum_{i \in B} (S_i^x \cos\phi - S_i^y \sin\phi).
\end{aligned} \qquad (3.33)$$

Changing to momentum space, this results in an additional term to the total Hamiltonian of the form

$$\mathcal{H}_{\mathbf{k}}^U = -(UM\cos\phi)\sigma^x \tau^0 - (UM\sin\phi)\sigma^y \tau^1 + UNM^2 \qquad (3.34)$$

which thus becomes

$$\mathcal{H}_{\mathbf{k}ss'} = \mathcal{H}_{\mathbf{k}ss'}^{\text{hop}} + \tilde{h}_0^x \sigma^x \tau^0 + \tilde{m}^y \sigma^y \tau^1 \qquad (3.35)$$

with $\tilde{m}^y = -UM\sin\phi$ and

$$\tilde{h}_0^x = h_0^x - UM\cos\phi. \qquad (3.36)$$

The eigenvalues of the Hamiltonian (3.35) read

$$\xi_{1\pm,\mathbf{k}} = \epsilon_{\mathbf{k}}^{(2)} - \sqrt{(\tilde{m}^y \pm g_{\mathbf{k}})^2 + (\tilde{h}_0^x \pm \epsilon_{\mathbf{k}}^{(1)})^2} - \mu,$$

$$\xi_{2\pm,\mathbf{k}} = \epsilon_{\mathbf{k}}^{(2)} + \sqrt{(\tilde{m}^y \pm g_{\mathbf{k}})^2 + (\tilde{h}_0^x \pm \epsilon_{\mathbf{k}}^{(1)})^2} - \mu.$$

Again, the self-consistency equations can be deduced from the grand-canonical potential and are given by

$$n = \frac{1}{N} \sum_{\alpha} \sum_{\beta=\pm} \sum_{\mathbf{k}} n_F(\xi_{\alpha\beta,\mathbf{k}}), \tag{3.37}$$

$$0 = \frac{UM}{N} \sum_{\alpha} \sum_{\beta=\pm} \sum_{\mathbf{k}} n_F(\xi_{\alpha\beta,\mathbf{k}}) \frac{\partial \xi_{\alpha\beta,\mathbf{k}}}{\partial \phi}, \tag{3.38}$$

$$M = \frac{1}{2N} \sum_{\alpha} \sum_{\beta=\pm} \sum_{\mathbf{k}} n_F(\xi_{\alpha\beta,\mathbf{k}}) \frac{\partial \xi_{\alpha\beta,\mathbf{k}}}{\partial M} \tag{3.39}$$

with

$$\frac{\partial \xi_{\alpha\beta,\mathbf{k}}}{\partial \phi} = (-1)^\alpha \frac{\mp g_{\mathbf{k}} \cos\phi - (h_0^x/2 \mp \epsilon_{\mathbf{k}}^{(1)}) \sin\phi}{\sqrt{(\tilde{m}^y \pm g_{\mathbf{k}})^2 + (\tilde{h}_0^x \pm \epsilon_{\mathbf{k}}^{(1)})^2}}, \tag{3.40}$$

$$\frac{\partial \xi_{\alpha\beta,\mathbf{k}}}{\partial M} = (-1)^\alpha \frac{(h_0^x/2 \pm \epsilon_{\mathbf{k}}^{(1)}) \cos\phi - UM \pm g_{\mathbf{k}} \sin\phi}{\sqrt{(\tilde{m}^y \pm g_{\mathbf{k}})^2 + (\tilde{h}_0^x \pm \epsilon_{\mathbf{k}}^{(1)})^2}}. \tag{3.41}$$

From Eqs. (3.38) and (3.40) it follows trivially that there is only a canted magnetization if there is a finite staggered SOC.

## 3.3 Metamagnetic Transition

Following Wohlfarth and Rhodes [48], a metamagnetic transition is expected in systems close to a magnetic instability. In the following, we therefore examine such a behavior for the two cases of fields applied in $z$ and $x$ direction. For this purpose, we first need to analyze the corresponding self-consistency equations in zero field.

## 3.3 Metamagnetic Transition

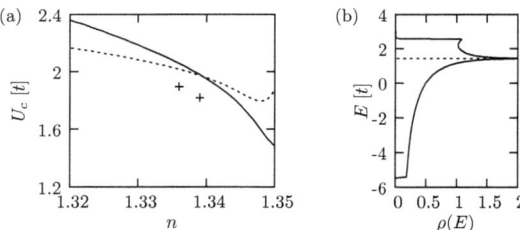

Figure 3.6: (a) Critical interaction strength for the ferromagnetic (solid line) and the SDW instability (dashed line). The crosses denote the choices for the interaction strength used for the two densities in Fig. 3.16. (b) Density of states in the absence of an external field showing the van Hove singularity at $\mu_{\rm vH} = 4t' = 1.44t$ (dashed line).

### 3.3.1 Field in $z$ direction

For a field applied perpendicular to the plane, we need to investigate the possible occurrence of an instability from Eq. (3.32) for the zero-field case. The linearized self-consistency equation yields the condition

$$1 - \frac{U}{N} \sum_{\alpha,\mathbf{k}} \frac{1}{4T\cosh^2(\xi_{\alpha\mathbf{k}s}/2T)} = 0 \quad (3.42)$$

for the occurrence of a ferromagnetic instability, the familiar Stoner criterion. This is not surprising since, apart from the folding of the Brillouin zone, the staggered SOC in the case of a magnetic field in $z$ direction only leads to a renormalization of the nn hopping [cf. Eq. (3.29)]. The critical interaction strength for a ferromagnetic instability to occur as a function of the electron density $n$ is shown by the solid line in Fig. 3.6(a). It drops significantly close to a density of $n_{\rm vH} \approx 1.35$. This corresponds to a chemical potential $\mu_{\rm vH} = 4t'$ where the Fermi surface hits the van Hove points located at $(\pm\pi, 0)$ and $(0, \pm\pi)$, thus leading to a diverging DOS. This divergence is shown in Fig. 3.6(b), where the density of states in zero field is plotted.

For a further analysis of the metamagnetic transition with applied field in $z$ direction, we fix the density of electrons slightly below $n_{\rm vH}$, $n = 1.336$, and

## Metamagnetism and Nematicity in $Sr_3Ru_2O_7$

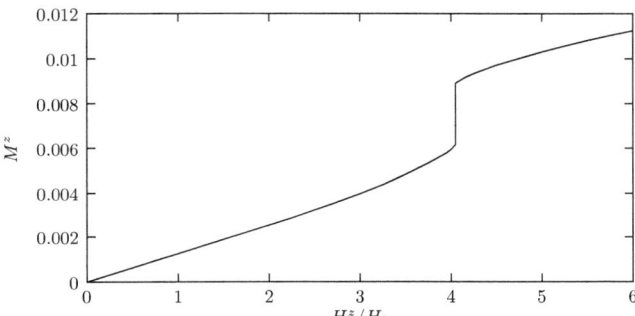

Figure 3.7: Magnetization for $n = 1.336$ for an applied field in $z$ direction for a temperature of $T = 5 \cdot 10^{-4} t$. The clear jump at the critical field $H_c$ is characteristic for a first-order transition.

choose an interaction strength $U$ close to the critical one obtained from the linearized self-consistency equation [see Fig. 3.6(a)]. Here and in following numerical calculations, we fix the nnn-hopping integral $t' = 0.36t$ and keep the spin-orbit-coupling strength at $\alpha = 0.05t$. The magnetization curve obtained is shown in Fig. 3.7. To emphasize the first-order nature of the transition at $T = 5 \cdot 10^{-4} t$, Fig. 3.8 shows the free energy as a function of applied magnetic field. Starting from the grand canonical potential (3.30), we perform a Legendre transformation and evaluate it as a function of $H^z$ with $n$ and $M^z$ calculated such as to minimize the resulting expression. We nicely obtain a kink in the free energy and the metastable states emerging that are characteristic for a first-order transition.

### 3.3.2 Field in $x$ direction

For the case of the field applied in $x$ direction, we can again explore the occurrence of a magnetic instability by linearizing the self-consistency equation (3.39)

## 3.3 Metamagnetic Transition

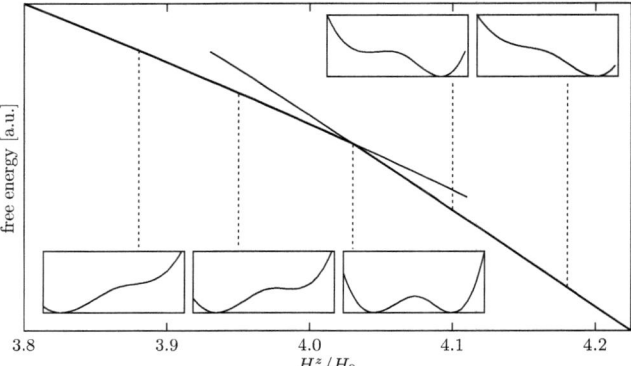

Figure 3.8: Free energy as a function of the magnetic field $H^z$. The two lines correspond to the low- and high-polarization phases and their crossing marks the critical field $H_c^z$ where the magnetization jumps. The thin lines show the metastable solutions that exist for magnetic fields close to the critical one. The insets show the free energy at fixed magnetic fields (denoted by the dashed lines) as a function of the magnetization $M^z$ between 0.005 and 0.01.

leading to

$$1 - \frac{U}{N} \Big\{ \sum_{\alpha,\beta} \sum_{\mathbf{k}} \frac{1}{4T \cosh^2(\xi_{\alpha\beta,\mathbf{k}}/2T)} \frac{(g_{\mathbf{k}} \sin\phi + \epsilon_{\mathbf{k}}^{(1)} \cos\phi)^2}{g_{\mathbf{k}}^2 + (\epsilon_{\mathbf{k}}^{(1)})^2}$$
$$+ \sum_{\alpha,\beta} \sum_{\mathbf{k}} n_{\mathrm{F}}(\xi_{\alpha\beta,\mathbf{k}})(-1)^{\alpha} \frac{(g_{\mathbf{k}} \cos\phi - \epsilon_{\mathbf{k}}^{(1)} \sin\phi)^2}{[g_{\mathbf{k}}^2 + (\epsilon_{\mathbf{k}}^{(1)})^2]^{3/2}} \Big\} = 0. \quad (3.43)$$

To further analyze this expression, it is useful to change to different order parameters, from $(M, \phi)$ to $(M^x = M\cos\phi, M^y = M\sin\phi)$ with $M^x$ the uniform magnetization in $x$ direction and $M^y$ the staggered component in $y$ direction, respectively. In these new parameters, the self-consistency equations read

$$M^x = \frac{1}{2N}\sum_{\alpha,\beta}\sum_{\mathbf{k}} n_F(\xi_{\alpha\beta,\mathbf{k}})\frac{(-1)^\alpha(\pm\epsilon_{\mathbf{k}}^{(1)} - UM^x)}{\sqrt{(\tilde{m}^y \pm g_{\mathbf{k}})^2 + (\tilde{h}_0^x \pm \epsilon_{\mathbf{k}}^{(1)})^2}}, \quad (3.44)$$

$$M^y = \frac{1}{2N}\sum_{\alpha,\beta}\sum_{\mathbf{k}} n_F(\xi_{\alpha\beta,\mathbf{k}})\frac{(-1)^\alpha(\pm g_{\mathbf{k}} - UM^y)}{\sqrt{(\tilde{m}^y \pm g_{\mathbf{k}})^2 + (\tilde{h}_0^x \pm \epsilon_{\mathbf{k}}^{(1)})^2}} \quad (3.45)$$

and, linearizing this system of equations, i.e.,

$$\begin{pmatrix} M^x \\ M^y \end{pmatrix} = \begin{pmatrix} \partial_x M^x & \partial_y M^x \\ \partial_x M^y & \partial_y M^y \end{pmatrix}\bigg|_{M^x=M^y=0} \begin{pmatrix} M^x \\ M^y \end{pmatrix}, \quad (3.46)$$

we obtain the conditions for two different possible magnetic instabilities,

$$0 = 1 - \frac{U}{N}\sum_{\alpha\mathbf{k}} \frac{1}{4T\cosh^2(\xi_{\alpha\beta,\mathbf{k}}/2T)}, \quad (3.47)$$

$$0 = \sqrt{t^2 + \alpha^2} + \frac{U}{N}\sum_{\alpha\mathbf{k}} \frac{(-1)^\alpha n_F(\xi_{\alpha\beta,\mathbf{k}})}{2|\cos k_x + \cos k_y|}. \quad (3.48)$$

The first one is the same as Eq. (3.42) and corresponds to a ferromagnetic instability. The second equation corresponds to a SDW instability occurring due to the near nesting of the Fermi surfaces. Note, however, that in both cases, the magnetization has a uniform as well as a staggered component. The solutions of these equations as functions of the electron filling $n$ are plotted in Fig. 3.6. The critical interaction strength for the SDW instability is generally below the one for a ferromagnetic instability, but shoots up when approaching the critical filling $n_{\text{vH}}$.

The linearized equation (3.43) can now be interpreted as having a ferromagnetic and a SDW contribution. Proximity to a SDW instability additionally lowers the critical interaction strength. Therefore, the metamagnetic transition could occur at a lower field then in the $z$-direction case, especially close to a SDW instability.

The magnetization due to a magnetic field applied in plane, as well as the amplitude of the staggered magnetization are plotted in Fig. 3.9. Again, we find a first-order transition for lower temperatures while the transition changes to a crossover upon increasing temperature. Note that the sign of $M^y$ depends on the sign of the SOC constant $\alpha$. There is no degeneracy in the state obtained which could lead to domain formation.

## 3.3 Metamagnetic Transition

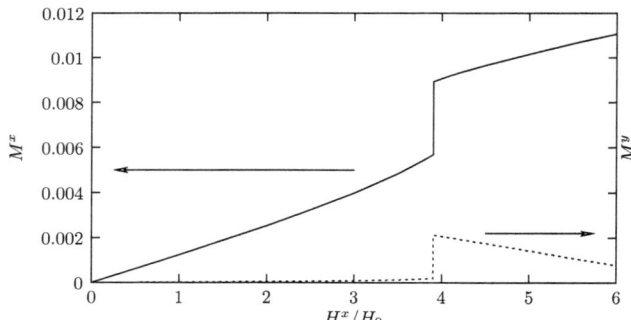

Figure 3.9: Uniform magnetization $M^x$ (solid line) and staggered magnetization $M^y$ (dashed line) for $n = 1.336$ for an applied field in $x$ direction for a temperature of $T = 5 \cdot 10^{-4} t$ and $\alpha = 0.05 t$. For this temperature, the metamagnetic transition is clearly of first order.

### 3.3.3 Comparison

Comparing the two cases of fields applied in $z$ and $x$ direction in Figs. 3.7 and 3.9, respectively, we first see that the critical field for the latter is shifted to lower fields, even though the zero-field susceptibilities differ by less then a percent, $\chi_0^z/\chi_0^x \approx 0.998$. This is due to the proximity of the system to a SDW instability as was pointed out above. This behavior is in qualitative agreement with the experimental phase diagram [see schematic phase diagram in Fig. 3.2(a)]. However, the difference of the in-plane and out-of-plane critical field is smaller in our model calculation than in the experiment. Our model only includes the staggered spin-orbit coupling entering through the oxygen displacement. Naturally, other SOC contributions, particularly from the Ru ions, would add to the anisotropy through an anisotropic $g$ tensor, likely with a larger polarizability in the basal plane than along the $z$ axis [58]. This is, however, beyond the scope of this study, as a detailed analysis would require to include other bands.

Second, a detailed study of the Gibbs free energy allows to determine the critical temperature $T^*$ up to which the metamagnetic transition is first order. For this purpose, we look at the magnitude of the magnetization jump $\Delta M$ as

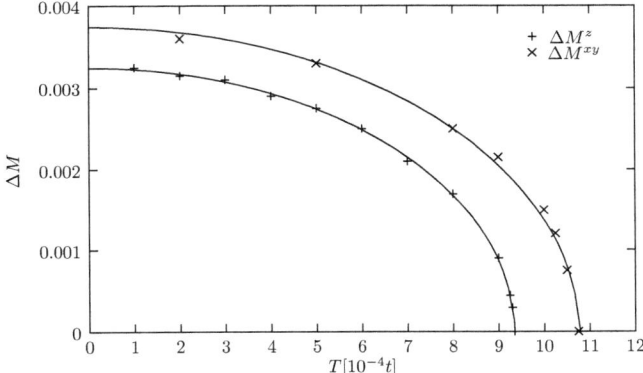

Figure 3.10: Jump in the magnetization at the metamagnetic transition for the two cases of a field applied in plane and out of plane. The vanishing of this jump signals the critical temperature above which the transition goes from being first-order to a crossover. Obviously, the critical temperature for the in-plane case is slightly higher.

a function of temperature. Fig. 3.10 shows the results obtained for both cases and for our choice of parameters we find $T_z^* \approx 9 \cdot 10^{-4} t$ while $T_x^* \approx 11 \cdot 10^{-4} t$. This anisotropy in the critical temperature is consistent with the trend in the experimental situation. However, it does not reproduce the quantum critical endpoint. Note that the difference between $T_z^*$ and $T_x^*$ could not be explained simply by an anisotropic $g$ tensor. In principle, it may be possible to tune the model in such a way as to press the critical temperature for out-of-plane fields $T_z^*$ to zero while still having a first-order transition at finite temperatures for in-plane fields. Also, fluctuation effects are likely important in this context. These features are, however, not essential to our discussion.

An additional important difference between in-plane and out-of-plane fields can be seen in Fig. 3.11, where the Fermi surfaces for both cases for fields below and above $H_c$ is shown. For both cases, the system undergoes a metamagnetic transition to prevent the majority-spin band from touching the van Hove points.

## 3.3 Metamagnetic Transition

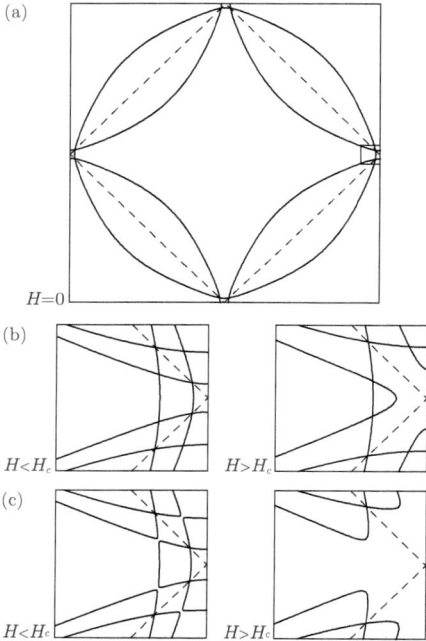

Figure 3.11: (a) Fermi surface for an electron density $n = 1.336$ without an applied field in the original Brillouin zone. Due to the rotation of the O octahedra, the bands are folded back with the new BZ denoted by the dashed lines. (b) and (c): Fermi surface just below and above the critical field in $z$ and $x$ direction, respectively. For clarity, only a small section of the BZ is shown, indicated by the little square around $(\pi, 0)$ in (a). For the case of a field applied in plane small gaps open close to the van Hove points even for $H < H_c$ (c).

In (c), we additionally see that the induced spin-density wave opens small gaps at the Fermi level close to the van Hove points. This has important consequences for the appearance of a nematic phase, as is discussed below.

## 3.4 Nematic Phase

In this section, we explore the occurrence of a nematic phase in our model for the two cases of a magnetic field applied in $z$ and $x$ direction, respectively. For this purpose, we introduce an additional interaction term [66, 69]

$$\mathcal{H}^n = \frac{1}{2N} \sum_{\mathbf{k},\mathbf{k}'} \sum_{s,s'} f_{\mathbf{k}\mathbf{k}'} n_{\mathbf{k}s} n_{\mathbf{k}'s'} \tag{3.49}$$

with a coupling function $f_{\mathbf{k}\mathbf{k}'}$ only contributing for zero momentum transfer, i.e., for forward scattering, which is the relevant interaction for a nematic phase to occur [70]. We then separate the coupling function according to

$$f_{\mathbf{k}\mathbf{k}'} = g d_{\mathbf{k}} d_{\mathbf{k}'} \tag{3.50}$$

with the forward-scattering strength $g$ and choose a $d_{x^2-y^2}$-symmetric form for the form factors, $d_{\mathbf{k}} = \cos k_x - \cos k_y$. This term can then lead to a nematic phase, reducing the symmetry from $C_4$ to $C_2$.

Introducing a mean-field decoupling for the (spin-independent) nematic order parameter

$$\eta = \frac{g}{N} \sum_{\mathbf{k}'} d_{\mathbf{k}'} \langle n_{\mathbf{k}'} \rangle, \tag{3.51}$$

we write for this interaction[1]

$$\mathcal{H}^n = \sum_{\mathbf{k}s} \eta d_{\mathbf{k}} n_{\mathbf{k}s} - \frac{N}{2g} \eta^2. \tag{3.52}$$

Since $d_{\mathbf{k}} = -d_{\mathbf{k}+\mathbf{Q}}$, but is isotropic in spin space, we can deal with it by replacing

$$\epsilon_{\mathbf{k}}^{(1)} \rightarrow \tilde{\epsilon}_{\mathbf{k}}^{(1)} = \epsilon_{\mathbf{k}}^{(1)} + \eta d_{\mathbf{k}}, \tag{3.53}$$

---

[1] Note that, as was shown by Adachi and Sigrist [71], on a mean-field level such a term can also be obtained from a nn repulsion without the need of introducing an engineered forward scattering with no clear microscopic origin.

## 3.4 Nematic Phase

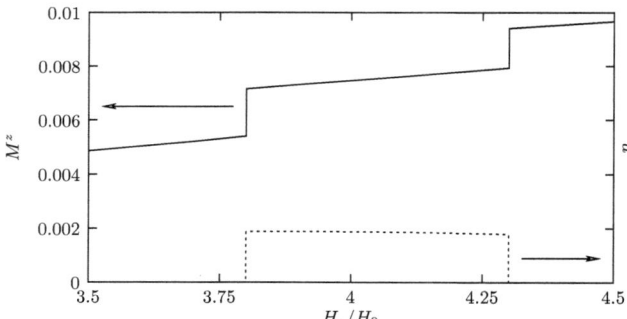

Figure 3.12: Mean-field results for the magnetization and the nematic order parameter $\eta$ for a field applied in $z$ direction for a temperature of $T = 5 \cdot 10^{-4} t$ showing an intermediate nematic phase bounded by two first-order transitions. Here, $g = 0.33$ and $n = 1.336$.

while all the above formulae still hold with the additional self-consistency equation

$$\eta = -\frac{g}{N} \sum_{\alpha=1,2} \sum_{\mathbf{k},s} n_\mathrm{F}(\xi_{\alpha\mathbf{k}s})(-1)^\alpha \frac{d_\mathbf{k} \tilde{\epsilon}_\mathbf{k}^{(1)}}{\sqrt{g_\mathbf{k}^2 + (\tilde{\epsilon}_\mathbf{k}^{(1)})^2}} \qquad (3.54)$$

for the $z$-direction case and

$$\eta = -\frac{g}{N} \sum_{\alpha,\beta} \sum_{\mathbf{k}} n_\mathrm{F}(\xi_{\alpha\beta,\mathbf{k}}) \frac{(-1)^\alpha d_\mathbf{k}(\tilde{\epsilon}_\mathbf{k}^{(1)} \pm \tilde{h}_0^x)}{\sqrt{(\tilde{m}^y \pm g_\mathbf{k})^2 + (\tilde{h}_0^x \pm \epsilon_\mathbf{k}^{(1)})^2}} \qquad (3.55)$$

for the $x$-direction case, respectively.

For sufficiently strong $g$ we find a magnetization curve for fields applied in $z$ direction as shown in Fig. 3.12. The two jumps in the magnetization border an intermediate phase with a finite value of the nematic order parameter $\eta$. The instability is in this case again driven mainly by electrons whose momenta lie close to the van Hove points. To obtain an intermediate phase before a single metamagnetic jump removes all such electrons from the Fermi surface, a critical scattering strength $g_c$ is thus necessary. For $g > g_c$, the nematic phase is entered

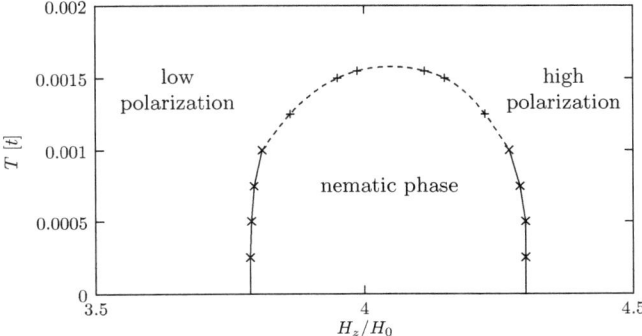

Figure 3.13: Phase diagram for a magnetic field applied in $z$ direction. While for low temperatures, the two consecutive transitions are of first order (solid line), they become second order before the nematic phase disappears completely (dashed line). Above this temperature, a metamagnetic crossover can still be seen.

from a low-polarization phase at some magnetic field $H_{c1}$ and left again towards a high-polarization phase at $H_{c2}$.

The $T$-$H$ diagram shown in Fig. 3.13 shows first-order transitions for low temperatures up to $T \approx 0.001t$, second-order transitions for higher temperatures until at $T \approx 0.0016t$ the nematic phase disappears completely to make way for a metamagnetic crossover (not shown). Similar behavior has already been observed in the calculations in Refs. [53, 72].

For the case of a field applied in $x$ direction, a very similar behavior is observed, with one important difference, however. Since the induced SDW already removes some weight from the Fermi surface close to the van Hove points before the critical field is reached, a larger forward-scattering strength is required for the occurrence of a nematic phase, i.e., $g_c^x > g_c^z$. Therefore, there is a range of $g$, where a nematic phase already exists for fields in $z$ direction, but only a single metamagnetic jump is observed for in-plane fields. This result is summarized in the phase diagram in Fig. 3.14. As a function of the SOC strength $\alpha$ and the forward-scattering strength $g$ we find three regimes: In addition to the two

## 3.5 Discussion and Conclusion

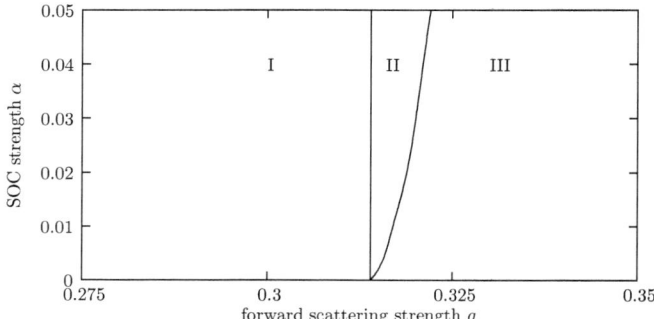

Figure 3.14: Critical forward-scattering strength for the cases of an applied field in $z$ and $x$ direction, respectively. We can distinguish three different regions. I: The forward-scattering strength is too week to enter a nematic phase, no matter in what direction the magnetic field is applied. Only a single metamagnetic transition occurs. II: While there is a nematic phase for fields applied in $z$ direction, no such phase occurs for fields in plane. This region corresponds to the case found in $Sr_3Ru_2O_7$. III: the forward-scattering strength is strong enough such that a nematic phase occurs for fields in any direction.

obvious regions, where there is either no intermediate phase at all (region I) or one for fields applied in any direction (III), there is now a new region with a nematic phase occurring only for fields applied in $z$ direction (II). Obviously, this is the region corresponding to the case of $Sr_3Ru_2O_7$.

## 3.5 Discussion and Conclusion

The range of the forward-scattering strength for which in our calculation a nematic phase is only observed for fields in $z$ direction is not very large. Note, however, that in $Sr_3Ru_2O_7$ the nematic phase only appears in a very narrow region $[(H_{c2} - H_{c1})/H_{c1} < 3\%$ [25]] and thus, $g$ is expected to be only slightly bigger than $g_c^z$. This is illustrated in Fig. 3.15, where the dependence of the critical fields on the forward-scattering strength is shown. The width of the nematic phase grows rather rapidly with increasing $g$. Therefore, the actual size of the forward-scattering strength might well lie in region II of Fig. 3.14.

Obviously, not only the strength of the SOC, but also the nesting properties of the spin-polarized Fermi surfaces play an important role for the appearance of the anisotropy effect. Nesting properties are a factor of tuning our model to the vicinity of a SDW instability. To analyze the impact of enhanced SDW correlations, we consider two different electron densities and corresponding on-site interaction strengths $U$ with $U_c^{FM} - U =$ const (see crosses in the left part of Fig. 3.6). This allows us to examine the cases of two different proximities to a SDW instability with comparable strengths of ferromagnetic correlations. One important finding is that the stronger the SDW correlations the more pronounced the anisotropy effect and thus, the smaller the ratio $H_c^x/H_c^z$ becomes. This is depicted in Fig. 3.16. In Fig. 3.6 we also show that the ferromagnetic and SDW instabilities can be competing for the chosen parameter range. We fix our model parameters in a way to avoid the occurrence of a staggered magnetic moment for any value of the magnetic field along the z axis, while the staggered moment is field induced for in-plane fields. For a more detailed discussion of the choice of parameters and the avoidance of an induced SDW for in-plane fields, see Appendix A.

One should also comment on the strength of the SOC $\alpha$ that can be expected for this system. To get an estimate of the on-site SOC strength $\lambda$ for $p$ electrons on the oxygen, we take the $O^{2-}$ vacuum value, $\lambda \sim 10\,\text{meV}$. A very crude estimate of the staggered SOC from Eq. (3.9) then yields

$$\frac{\alpha}{t} \approx \frac{2\lambda}{\Delta} \frac{t'_{pd}}{t_{pd}} \approx \frac{2\lambda}{\Delta}. \tag{3.56}$$

Estimating $\Delta \approx 1.5\,\text{eV}$, we find that for $\alpha$ a value on the order of a percent of $t$ seems reasonable [37]. Comparing Figs. 3.14 and 3.15, this would allow for a nematic phase with a width of $(H_{c2} - H_{c1})/H_{c1} \approx 2\%$, in agreement with experiment. For a more reliable estimate of $\alpha$, DFT calculations should be performed.

Finally, some general remarks on the nematic phase are in order. As was already mentioned, the nematic phase introduced here is the same as discussed by other authors [53, 54, 55]. As shown by these authors, the nematic phase could account for several experimentally observed phenomena like the anomalous

## 3.5 Discussion and Conclusion

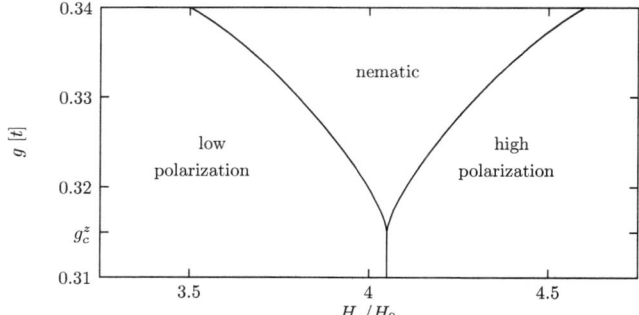

Figure 3.15: Phase diagram for a magnetic field applied in $z$ direction depending on the forward scattering strength $g$. Below a critical strength $g_c^z$, no nematic phase is entered, but the system undergoes a single metamagnetic transition connected to the proximity to the ferromagnetic instability as discussed above. Above $g_c^z$, the system enters a nematic phase whose region grows with increasing $g$. Here, $T = 5 \cdot 10^{-4} t$.

resistivity, or the non-Fermi-liquid behavior of the susceptibility and the specific-heat coefficient.

In conclusion, we showed in this chapter that the rotated oxygen octahedra lead to a staggered spin-dependent hopping that can be described with the help of a (staggered) SOC of Rashba type. This introduces an anisotropy of the response to a magnetic field, namely an induced spin-density wave for the case of in-plane fields. This staggered magnetization could be observed in neutron-scattering experiments. To our knowledge, this kind of experiment has not been performed so far. The additional magnetization has, first, the effect that the critical field for a metamagnetic transition is shifted to lower values for in-plane fields. Also, the critical temperature $T^*$ up to which the transition is first order is higher for fields in the $xy$ plane. Last and most important, the SDW opens gaps at the Fermi level that lead to an anisotropy for the appearance of a nematic phase. Additionally considering SOC effects of the Ru orbital would account for the full anisotropy of $H_c$ ($g$-tensor anisotropy). Therefore, the work presented in

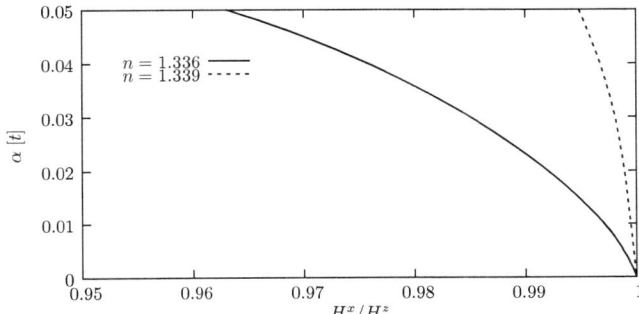

Figure 3.16: Anisotropy of the critical field in $x$ direction compared to the value for the $z$ direction for the densities $n = 1.336$ (solid line) and $n = 1.339$ (dashed line), respectively. The value for the on-site interaction is chosen as indicated in Fig. 3.6. The dependence on the strength of the spin-orbit coupling is more pronounced the closer the system is to a SDW instability.

this chapter allows for a picture that is qualitatively consistent with experimental observations including the anisotropies in $H_c$, $T^*$ and the appearance of a nematic phase.

## 3.5 Discussion and Conclusion

# Chapter 4

# Superconductivity in Crystals with Locally Broken Inversion Symmetry

> *Symmetry aspects play a crucial role in the study of superconductivity, especially for the determination of possible pairing channels. In this chapter the analysis of lacking inversion symmetry is generalized to the case of local reduction by simultaneous global invariance. Motivated by the crystal structure in the preceding chapter and the SOC as a result thereof, two examples are studied, the case of symmetry lacking bonds and planes, respectively. In addition, comparison to the case of antiferromagnetic superconductors, i.e., systems with locally broken time-reversal symmetry, is made.*

## 4.1 Introduction

Superconductivity is one of the most intriguing states of matter and almost 100 years after its discovery still one of the main drivers of innovation in both, theoretical and experimental condensed matter physics. It took almost 50 years until Bardeen, Cooper, and Schrieffer proposed their famous theory realizing that the superconducting state has to be described based on an instability of the Fermi

## 4.1 Introduction

sea towards the formation of so-called Cooper pairs [73]. In their original theory, these are electron pairs with opposite momentum and spin with energies very close to the Fermi energy, forming a bound pair with a spin-singlet configuration. As first realized by Anderson in 1959, the existence of such degenerate electron pairs is guaranteed by time-reversal symmetry which relates the states $|\mathbf{k},\uparrow\rangle$ with $|-\mathbf{k},\downarrow\rangle$ and thus implies $\xi_{\mathbf{k}\uparrow} = \xi_{-\mathbf{k}\downarrow}$ (Anderson's first theorem) [74]. Therefore, breaking of this symmetry by an external magnetic field or magnetic impurities substantially suppresses superconductivity. However, Baltensperger and Straessler showed that superconductivity and antiferromagnetism, i.e., the local breaking of time-reversal symmetry periodically as to retain the global symmetry, can coexist [75]. Long known examples are magnetic superconductors, e.g., the ternary rare earth compounds in the Chevrel phase $REMo_6S_8$ (RE = Gd, Tb, Dy and Er), where the localized moments on the rare earth ions order antiferromagnetically and, still, the Fermi sea undergoes a superconducting transition [76]. Interestingly, there were also examples of coexistence found in strongly correlated electron systems, where the same electrons carry the magnetic moment and participate in the superconductivity as in $CeIn_3$ [77], $CeRhIn_5$ [78, 79] or recently in the spin-ladder system $Sr_{14-x}Ca_xCu_{24}O_{41}$ [80].

For a pairing of electrons in a spin-triplet configuration, Anderson noticed later that an additional symmetry has to be present, namely inversion symmetry (Anderson's second theorem) [26]. The discovery of superconductivity in crystals lacking an inversion center ("non-centrosymmetric superconductivity") and yet showing signs of triplet pairing therefore attracted much attention in recent years and these systems have been intensively studied [81, 82]. On the level of a microscopic description, the lack of inversion symmetry manifests itself in a symmetry-reducing term in the Hamiltonian. This term, that in $CePt_3Si$ has the form of an antisymmetric SOC of Rashba type, couples the spin-singlet with the spin-triplet gap. In addition, the direction of the $d$ vector describing the spin-triplet gap becomes fixed even for small antisymmetric SOC.

As discussed above, systems lacking time-reversal symmetry globally and locally - in other words with both, 'ferro-type' and 'antiferro-type' order - have been intensively studied. However, the analysis of inversion-symmetry breaking has mainly been focused on crystals globally lacking such invariance. Recently, Yanase has studied the case of locally broken inversion symmetry due to stacking faults, where the global inversion symmetry is retained because of the random distribution of these faults [83]. Motivated by the special crystal structure of $Sr_3Ru_2O_7$, where the inversion symmetry on the Ru-Ru bonds is broken without the system lacking global inversion symmetry and the staggered SOC as a consequence thereof, we study the effect of broken inversion symmetry with an order of antiferro type. For this purpose, we look at two examples, the first of which is the square lattice with a distortion as described in chapter 3. The second example we want to study consists of layers where, for each single layer, the mirror symmetry on the $xy$ plane is broken (see top left of Fig. 4.1). The global inversion symmetry is then recovered by an alternating stacking of these layers. One of the key characteristics of these systems is that they can be split into two sublattices and we, thus, have to implement a formulation based on a two-band model. This also allows to make a connection to systems with an antiferromagnetic background. Note, however that we are only interested in the possible symmetry-allowed gap couplings and the arising consequences and not in the effects on possible mechanisms mediating superconductivity.

In order to be able to analyze superconductivity in the above mentioned crystals, we first recall the basic theory of superconductivity in the next section. The two case studies then follow in sections 4.3 and 4.4. As symmetry and associated group theoretical considerations have proven to be of considerable use in the study of symmetry-reduced systems, each section is started with a symmetry analysis before turning to the microscopic description of the gap equation. Finally the chapter ends with a discussion and comparison of the two cases.

## 4.2 Basic Theory of Superconductivity

### 4.2.1 Green's functions and Gor'kov equations

In this section, a brief overview over the formulation of the theory of superconductivity is given in terms of Green's functions leading to the Gor'kov equations. For this purpose, we consider a general Hamiltonian of the form

$$\mathcal{H} = \mathcal{H}_0 + \mathcal{H}', \tag{4.1}$$

## 4.2 Basic Theory of Superconductivity

consisting of a non-interacting part

$$\mathcal{H}_0 = \sum_{\mathbf{k},s} \xi_{\mathbf{k}s} c^\dagger_{\mathbf{k}s} c_{\mathbf{k}s} \qquad (4.2)$$

and a pairing interaction

$$\mathcal{H}' = \frac{1}{N} \sum_{\mathbf{k},\mathbf{k}',\mathbf{q}} \sum_{s_1 \cdots s_4} V^{s_1 s_2 s_3 s_4}(\mathbf{k},\mathbf{k}') c^\dagger_{\mathbf{k}s_1} c^\dagger_{-\mathbf{k}+\mathbf{q}s_2} c_{-\mathbf{k}'+\mathbf{q}s_3} c_{\mathbf{k}'s_4}. \qquad (4.3)$$

We define the electron Green's function for finite temperatures $T$ as

$$G^{ss'}(\mathbf{k},\tau) = -\langle T_\tau c_{\mathbf{k}s}(\tau) c^\dagger_{\mathbf{k}s}(0) \rangle, \qquad (4.4)$$

where $T_\tau$ is the time-ordering operator with respect to the imaginary time $\tau$ and the angular brackets denote thermal averaging with respect to the Hamiltonian (4.1). In frequency-momentum space we then have

$$G^{ss'}(\mathbf{k},\tau) = T \sum_n G^{ss'}(\mathbf{k},\omega_n) e^{-i\omega_n \tau} \qquad (4.5)$$

with the fermionic Matsubara frequencies $\omega_n = (2\pi+1)nT$. This Green's function can now be calculated using the equation of motion for Heisenberg operators,

$$\frac{\partial c_{\mathbf{k}s}(\tau)}{\partial \tau} = [\mathcal{H}, c_{\mathbf{k}s}(\tau)]. \qquad (4.6)$$

Introducing the two so-called anomalous Green's functions,

$$F^{ss'}(\mathbf{k},\tau) = -\langle T_\tau c_{\mathbf{k}s}(\tau) c_{-\mathbf{k}s'}(0) \rangle, \qquad (4.7)$$
$$F^{\dagger ss'}(\mathbf{k},\tau) = -\langle T_\tau c^\dagger_{\mathbf{k}s}(\tau) c^\dagger_{-\mathbf{k}s'}(0) \rangle, \qquad (4.8)$$

we find a set of equations,

$$\{i\omega_n - \xi_{\mathbf{k}s}\} G(\mathbf{k},\omega_n) + \Delta(\mathbf{k}) F^\dagger(\mathbf{k},\omega_n) = \sigma^0, \qquad (4.9)$$
$$\{i\omega_n + \xi_{\mathbf{k}s}\} F^\dagger(\mathbf{k},\omega_n) - \Delta^\dagger(\mathbf{k}) G(\mathbf{k},\omega_n) = 0, \qquad (4.10)$$
$$\{i\omega_n - \xi_{\mathbf{k}s}\} F(\mathbf{k},\omega_n) - \Delta(\mathbf{k}) G^T(\mathbf{k},\omega_n) = 0, \qquad (4.11)$$

the Gor'kov equations. Here, we have introduced the gap function

$$\Delta^{ss'}(\mathbf{k}) = -T \sum_{n,\mathbf{k}'} \sum_{s_3,s_4} V^{ss' s_3 s_4}(\mathbf{k},\mathbf{k}') F^{s_4 s_3}(\mathbf{k}',\omega_n) \qquad (4.12)$$

and $\sigma^0$ is the $(2 \times 2)$ identity matrix. Note that Eqs. (4.9)-(4.11) are matrix equations since all the above defined Green's functions as well as the gap function are $2 \times 2$ matrices. With the help of Eq. (4.11), we can now write the self-consistency equation for the gap function,

$$\Delta^{ss'}(\mathbf{k}) = -T \sum_n \sum_{\mathbf{k'}} \sum_{s_3,s_4} V^{ss's_3s_4}(\mathbf{k},\mathbf{k'})$$
$$\times [G_0(\mathbf{k'},\omega_n)\Delta(\mathbf{k'})G^T(-\mathbf{k'},-\omega_n)]^{s_4s_3}, \quad (4.13)$$

where $G_0(\mathbf{k},\omega_n) = (i\omega_n - \xi_{\mathbf{k}s})^{-1}$ is the Green's function of the non-interacting system described by $\mathcal{H}_0$.

### 4.2.2 Crystal symmetry aspects

The gap function is the superconducting order parameter and for the characterization of the superconducting state we thus need to analyze it. However, Eq. (4.13) is a complicated functional equation whose solutions in general have to be found numerically. The aim of a symmetry analysis is to decompose the space of gap functions into sectors that do not mix through the gap equation (4.13), therefore simplifying the problem. It also allows us to determine which solutions are connected in terms of symmetry and therefore are degenerate. In addition, knowledge of the superconducting gap symmetry might give us valuable hints on the microscopic origin of the superconductivity of the system.

The symmetry properties of the gap are determined by the form of the interaction and the total symmetry of the system. First, since the gap is a $2 \times 2$ matrix, we can write it as

$$\Delta(\mathbf{k}) = \begin{pmatrix} \Delta^{\uparrow\uparrow}(\mathbf{k}) & \Delta^{\uparrow\downarrow}(\mathbf{k}) \\ \Delta^{\downarrow\uparrow}(\mathbf{k}) & \Delta^{\downarrow\downarrow}(\mathbf{k}) \end{pmatrix} = \sum_{i=0}^{3} \Delta_i(\mathbf{k})\sigma^i, \quad (4.14)$$

where, for the last equality, we have used the fact that the Pauli matrices together with $\sigma^0$ form a basis for $M_2(\mathbb{C})$, the space of $2 \times 2$ matrices. Due to Pauli's principle, the gap has to fulfill $\Delta^{ss'}(\mathbf{k}) = -\Delta^{s's}(-\mathbf{k})$.[1] Therefore, it follows that

---

[1] In detail, because of $V^{ss's_3s_4}(\mathbf{k},\mathbf{k'}) = \langle \mathbf{k},s;-\mathbf{k},s'|V|-\mathbf{k'},s';\mathbf{k'},s\rangle$ and the fermionic anticommutation rule.

## 4.2 Basic Theory of Superconductivity

$\Delta_2(\mathbf{k})$ in Eq. (4.14) has to be an even function in $\mathbf{k}$ while all the others need to be odd in $\mathbf{k}$. This is usually written in the literature in the form

$$\Delta(\mathbf{k}) = (\psi(\mathbf{k})\sigma^0 + \vec{d}(\mathbf{k}) \cdot \vec{\sigma})i\sigma^y = \psi(\mathbf{k})\varsigma^0 + \vec{d}(\mathbf{k}) \cdot \vec{\varsigma}, \qquad (4.15)$$

where we have introduced $\varsigma^0 = i\sigma^y$ and $\vec{\varsigma} = \vec{\sigma}i\sigma^y$ for simplicity of notation. In Eq. (4.15), we have therefore split the gap into a spin-singlet part with $\psi(-\mathbf{k}) = \psi(\mathbf{k})$ and a spin-triplet part with $\vec{d}(-\mathbf{k}) = -\vec{d}(\mathbf{k})$.

Second, since the gap has to reflect the symmetry of the system, we can expand the gap in basis functions of the irreducible representation (IR) of the symmetry group of the crystal. For the simplest case of a system with inversion symmetry, the IRs can be divided into even and odd ones resulting in a gap,

$$\Delta(\mathbf{k}) = \sum_a \Delta_a \psi_a(\mathbf{k})\varsigma^0 + \sum_i \sum_b \Delta_b^i d_b^i(\mathbf{k})\varsigma^i, \qquad (4.16)$$

where the $\psi_a(\mathbf{k})$ and $d_b^i(\mathbf{k})$ are normalized even and odd basis functions of the IRs, respectively. In this case, even and odd gap functions do not mix and, likewise, spin-singlet and spin-triplet gap functions are decoupled. If, in addition, there is no or only a small SOC, the full symmetry Group $\mathcal{G}$ includes both, the space group of the crystal $\mathcal{G}_0$ and the group of spin rotations $\mathcal{G}_s$, $\mathcal{G} = \mathcal{G}_0 \times \mathcal{G}_s$. This means that spin and momentum degrees of freedom are decoupled, i.e., the direction of the $d$ vector is not fixed by the momentum dependence of the gap. As an example, looking at table B.2 for the group $D_{4h}$, the odd functions of momentum in the left column of basis functions can point in any direction in spin space.

So far, we have assumed the simplest case of a crystal with inversion symmetry and no SOC. If these conditions are violated, this has important consequences for the superconducting state:

- For systems with strong SOC, the spin and momentum degrees of freedom couple such that the full symmetry group only consists of the space group of the crystal, $\mathcal{G} = \mathcal{G}_0$. Therefore, spin rotations are tied to orbital rotations, resulting in basis functions composed of both, a momentum and a spin part (cf. table B.2).

- If the symmetry group of the system does not include inversion, it is not possible to find even and odd eigenfunctions anymore. As a consequence, even (spin-singlet) and odd (spin-triplet) gaps are not independent anymore but couple through the gap equation (4.13).

### 4.2.3 Linearized gap equation

To analyze the instability condition for superconductivity and the possible gap couplings, we linearize the gap equation by using Eq. (4.11), which in a first approximation can be written as

$$F(\mathbf{k},\omega_n) = G_0(\mathbf{k},\omega_n)\Delta(\mathbf{k})G_0^T(-\mathbf{k},-\omega_n). \tag{4.17}$$

The linearized gap equation thus yields

$$\Delta^{ss'}(\mathbf{k}) = -T \sum_n \sum_{\mathbf{k}'} V^{ss's_3s_4}(\mathbf{k},\mathbf{k}')[G_0(\mathbf{k}',\omega_n)\Delta(\mathbf{k}')G_0^T(-\mathbf{k}',-\omega_n)]^{s_4s_3}, \tag{4.18}$$

where we adopted the convention of summing over multiply appearing (spin) indices. We now parametrize the pairing interaction in a similar way as the gap function,

$$V^{s_1s_2s_3s_4}(\mathbf{k},\mathbf{k}') = \sum_a v_a\,[\psi_a(\mathbf{k})\varsigma^0]_{s_1s_2}\,[\psi_a(\mathbf{k}')\varsigma^0]^\dagger_{s_3s_4}$$
$$+ \sum_{i,b} v_b^i\,[d_b^i(\mathbf{k})\varsigma^i]_{s_1s_2}\,[d_b^i(\mathbf{k}')\varsigma^i]^\dagger_{s_3s_4}, \tag{4.19}$$

where we again expanded in basis functions of the IRs. Using

$$\frac{1}{2}\mathrm{Sp}[(\varsigma^i)^\dagger \varsigma^j] = \delta_{ij} \tag{4.20}$$

and the orthogonality of the different basis functions, we find for the singlet gap

$$1 = -\frac{v_a}{\beta}\sum_n \sum_{\mathbf{k}'} \psi_a^*(\mathbf{k}')G_0(\mathbf{k}',\omega_n)\psi_a(\mathbf{k}')G_0^T(-\mathbf{k}',-\omega_n) \tag{4.21}$$

and analogously for the spin-triplet case. These are self-consistency conditions for the critical temperature $T_c$ for superconductivity for every channel, where the channel with the highest critical temperature yields the leading instability.

Note that had we not chosen basis functions corresponding to the correct symmetry, Eq. (4.21) would not be diagonal, but different basis functions $\psi_a(\mathbf{k})$ and

### 4.3 Inversion-Symmetry Lacking Bonds

$d_b^i(\mathbf{k})$ would actually mix through the non-interacting Green's function as we encounter in the next sections. Also note, however, that only gap symmetries that actually appear in the expansion of the interaction are possible. Therefore, even though there might be symmetry-allowed couplings of different gap functions, they are only relevant if the respective pairing channels are actually supported by the dominant pairing interaction. For the analysis of the possible gap couplings we thus first examine the symmetry of the system through the non-interacting Hamiltonian. In a next step, the analysis of the pairing potential then yields the actually appearing couplings. Last, the leading instability can be found by (numerical) evaluation of the linearized gap equation (4.18).

## 4.3 Inversion-Symmetry Lacking Bonds

The first example analyzed is motivated by the crystal structure of $Sr_3Ru_2O_7$ as introduced in chapter 3. We again focus only on a single $RuO_4$ layer with the O octahedra rotated around the $c$ axis as shown in Fig. 3.4. As was already discussed in the previous chapter, this shifts the bond oxygens to off-center positions, thus breaking inversion symmetry on the bonds. Due to the antiferro-type arrangement, the crystal retains global inversion symmetry. After a brief symmetry analysis of such a layer, we examine in the following the influence of this lattice distortion on the possible superconducting pairing channels and gap couplings.

### 4.3.1 Symmetry classification

Without the rotation of the octahedra, the factor group $\mathcal{G}/T$ with $\mathcal{G}$ the space group and $T$ the translation group of the $RuO_4$ layer, respectively, contains the following elements: the identity $E$, a four-fold rotation $C_4$ around $z$, two two-fold axes $C_2'$ around $x$ and $y$, and another two two-fold axes $C_2''$ around the in-plane diagonals. In addition, the system also possesses inversion symmetry. The factor group $\mathcal{G}/T$ of such a single layer is thus $D_{4h}$ and the crystal is symmorphic.

We are now interested in the situation where the O octahedra are rotated and the unit cell is doubled as depicted by the green and yellow lattice sites in Fig. 3.4. The factor group $\mathcal{G}/T$ of this crystal contains the elements $\{E|0\}$,

## Superconductivity in Crystals with Locally Broken Inversion Symmetry

|  | intra-sublattice | inter-sublattice | $A \leftrightarrow B$ | IR |
|---|---|---|---|---|
| intra-band | $\tau^0$ | $\tau^3$ | + | $A_{1g}$ |
| inter-band | $\tau^1$ | $\tau^2$ | − | $A_{2g}$ |

Table 4.1: The different terms possible in a Hamiltonian in momentum representation. Since an interchange of the lattice sites $A$ and $B$ corresponds to a rotation $C_2'$ or $C_2''$, the terms containing $\tau^0$ and $\tau^3$ belong to the IR $A_{1g}$, while $\tau^1$ and $\tau^2$ belong to $A_{2g}$.

$\{C_4|0\}$, $\{C_4^2|0\}$, $\{C_4^3|0\}$, $\{C_2'|\vec{a}\}$, $\{C_2''|\vec{a}\}$ as well as the inversion. Here, the notation $\{R|\vec{r}\}$ describes a rotation $R$ accompanied by a translation by the vector $\vec{r}$ which is not a primitive translation of the Bravais lattice and $\vec{a}$ is a lattice vector of the former lattice. The factor group $\mathcal{G}/T$ is thus still isomorphic to $D_{4h}$, with $C_2'$ and $C_2''$, however, combined with a shift of the lattice, i.e., an interchange of the two sublattices. The crystal under consideration is therefore non-symmorphic.

Regarding the Hamiltonian, this means that there are now terms allowed that were formerly forbidden by symmetry. For a further analysis, we again note that the bipartite lattice leads to two electron bands which, in momentum space, can be described with the help of Pauli matrices $(\tau^0, \vec{\tau})$ for electrons with momentum $\mathbf{k}$ and $\mathbf{k}+\mathbf{Q}$ (cf. chapter 3). These four matrices correspond to four different types of terms that can appear in the momentum representation of the Hamiltonian. As we saw in the last chapter in Eq. (3.12), a simple nn hopping (connecting sublattices) can be written with the help of $\tau^3$, while a nnn hopping or the chemical potential term (within the sublattices) correspond to $\tau^0$. These two intra-band terms do not change sign when the two sublattices are interchanged, $A \leftrightarrow B$. On the other hand, the staggered SOC coming from Eq. (3.11), again an inter-sublattice term, changes sign under sublattice interchange and corresponds to the inter-band matrix $\tau^2$. As a last possibility, a staggered magnetization is an intra-sublattice term changing sign under interchange of $A$ and $B$ and yields the inter-band matrix $\tau^1$ as we saw in Eq. (3.35). The sign under sublattice interchange also determines the IR the term belongs to, because an interchange of sublattices has the same effect as the rotations $C_2'$ and $C_2''$. These findings are summarized in Table. 4.1. The IR for the Pauli spin matrices follow from the fact

## 4.3 Inversion-Symmetry Lacking Bonds

that the $\sigma^i$ are representations of the rotation group. For $D_{4h}$, we thus find that $\sigma^z$ belongs to $A_{2g}$, while $\sigma^x$ and $\sigma^y$ belong to $E_g$. Finally, the spin symmetric $\sigma^0$ belongs to $A_{1g}$.

Since the Hamiltonian $\mathcal{H}_0$ has to be invariant under the operations of the symmetry group of the crystal, its terms have to belong to $A_{1g}$. However, with the special structure of spin and orbital indices, we can have terms of the form $(A_{2g} \otimes A_{2g})$ which do belong to $A_{1g}$. This allows for terms of the form

$$\mathcal{H}_\mathbf{k}^{\text{soc}} = g_\mathbf{k} \sigma^z \otimes \tau^2, \tag{4.22}$$

where $g_\mathbf{k} = 2\alpha(\cos k_x + \cos k_y)$ is a nn-hopping term with $A_{1g}$ symmetry. This is the term we derived microscopically in Chapter 3. Note that, here, we have written the product of the two Pauli matrices explicitly as a tensor product to emphasize that the matrices are acting on spin and orbital space, respectively.

### 4.3.2 Symmetry-allowed gap couplings

Analogously to the expansion of the gap in Eq. (4.14) for the case of a single band, we parametrize the gap in terms of tensor products of spin and orbital Pauli matrices,

$$\Delta_{\alpha\beta}^{ss'}(\mathbf{k}) = \sum_{i,j} \Delta_{ij}(\mathbf{k})(\varsigma_{ss'}^i \otimes \tau_{\alpha\beta}^j). \tag{4.23}$$

Again, the parity of the momentum dependent part $\Delta_{ij}(\mathbf{k})$ depends on the sign of $\varsigma^i \otimes \tau^j$ under the interchange of the index pair $(s, \alpha) \leftrightarrow (s', \beta)$ and can be expanded as $\Delta_{ij}(\mathbf{k}) = \sum_a \Delta_{ij}^{(a)} \psi_{ij}^{(a)}(\mathbf{k})$. In this expansion, $\psi_{ij}^{(a)}(\mathbf{k})$ is a basis function of the correct parity. Note that distinguishing spin-singlet and spin-triplet gap functions is not applicable anymore, but due to global inversion symmetry a distinction in even and odd momentum dependence should be made instead. The system's invariance under inversion then still guarantees that the spaces of even and odd gap functions are decoupled.

We now want to examine which terms in the expansion (4.23) can couple to each other. As seen in the previous section, all the elements of the Hamiltonian belong to the IRs $(A_{1g} \otimes A_{1g})$ and $(A_{2g} \otimes A_{2g})$, respectively. For a coupling to occur in lowest order between a gap of symmetry $(R_m \otimes R_n)$ to one of $(R_i \otimes R_j)$, the latter has to appear in the decomposition of $(R_m \otimes R_n)(A_{2g} \otimes A_{2g})$. For the

*Superconductivity in Crystals with Locally Broken Inversion Symmetry*

spin part of the gap, we note that $\varsigma^x$ and $\varsigma^y$ belong to $E_g$ (the triplet part $d^x$ and $d^y$), $\varsigma^z$ belongs to $A_{2g}$ (the triplet part $d^z$) while $\varsigma^0$ belongs to $A_{1g}$ (the singlet part). For the orbital part, we can use table 4.1 to find the following symmetry-allowed gap couplings,

$$(A_{2g} \otimes A_{2g}) \leftrightarrow (A_{1g} \otimes A_{1g}), \tag{4.24}$$

$$(A_{2g} \otimes A_{1g}) \leftrightarrow (A_{1g} \otimes A_{2g}), \tag{4.25}$$

$$(E_g \otimes A_{2g}) \leftrightarrow (E_g \otimes A_{1g}). \tag{4.26}$$

For a detailed list of gap couplings, see Appendix C.

### 4.3.3 Microscopic consideration of gap-function coupling

The general scheme applied to analyze the superconducting instability and the possible gap couplings follows the one introduced in section 4.2. Note, however, that now all the Green's functions as well as the gap function are $4 \times 4$ matrices in spin and orbital space. The linearized gap equation reads

$$\Delta_{\alpha\beta}^{ss'}(\mathbf{k}) = -T \sum_n \sum_{\mathbf{k}'} V_{\alpha\beta,\mu\nu}^{ss's_3s_4}(\mathbf{k},\mathbf{k}')[G_0(\mathbf{k}',\omega_n)\Delta(\mathbf{k}')G_0^T(-\mathbf{k}',-\omega_n)]_{\nu\mu}^{s_4s_3}. \tag{4.27}$$

We now have to choose a specific Hamiltonian with a pairing interaction. For the non-interacting part we consider the hopping Hamiltonian (3.14) with (3.15) derived in chapter 3,

$$\mathcal{H}^{\text{hop}} = \sum_{\mathbf{k}} \vec{c}_{\mathbf{k}}^{\dagger} \mathcal{H}_{\mathbf{k}}^{\text{hop}} \vec{c}_{\mathbf{k}} \tag{4.28}$$

with $\vec{c}_{\mathbf{k}}^{\dagger} = (c_{\mathbf{k}\uparrow}^{\dagger}, c_{\mathbf{k}\downarrow}^{\dagger}, c_{\mathbf{k}+\mathbf{Q}\uparrow}^{\dagger}, c_{\mathbf{k}+\mathbf{Q}\downarrow}^{\dagger}) \equiv (c_{1\mathbf{k}\uparrow}^{\dagger}, c_{1\mathbf{k}\downarrow}^{\dagger}, c_{2\mathbf{k}\uparrow}^{\dagger}, c_{2\mathbf{k}\downarrow}^{\dagger})$ and

$$\mathcal{H}_{\mathbf{k}}^{\text{hop}} = (\varepsilon_{+,\mathbf{k}} - \mu)\sigma^0 \otimes \tau^0 + \varepsilon_{-,\mathbf{k}}\sigma^0 \otimes \tau^3 - g_{\mathbf{k}}\sigma^z \otimes \tau^2. \tag{4.29}$$

Here, we have chosen a slightly more general formulation, where $\varepsilon_{+,\mathbf{k}}$ denotes intra-sublattice hopping ($\varepsilon_{+,\mathbf{k}+\mathbf{Q}} = \varepsilon_{+,\mathbf{k}}$) and $\varepsilon_{-,\mathbf{k}}$ denotes inter-sublattice hopping ($\varepsilon_{-,\mathbf{k}+\mathbf{Q}} = -\varepsilon_{-,\mathbf{k}}$), both of the undistorted lattice. The corresponding non-interacting Green's functions can straightforwardly be calculated by inverting the ($4 \times 4$) matrix ($i\omega_n \sigma^0 \otimes \tau^0 - \mathcal{H}_{\mathbf{k}}^{\text{hop}}$) yielding

$$G_0(\mathbf{k}, \omega_n) = G_{0+}(\mathbf{k}, \omega_n)\sigma^0 \otimes \tau^0 - G_{0-}(\mathbf{k}, \omega_n)(\hat{g}_{\mathbf{k}}\sigma^z \otimes \tau^2 - \hat{\varepsilon}_{-,\mathbf{k}}\sigma^0 \otimes \tau^3), \tag{4.30}$$

## 4.3 Inversion-Symmetry Lacking Bonds

where
$$G_{0\pm}(\mathbf{k},\omega_n) = \frac{1}{2}\left(\frac{1}{i\omega_n - \xi_{+,\mathbf{k}}} \pm \frac{1}{i\omega_n - \xi_{-,\mathbf{k}}}\right), \qquad (4.31)$$

$\hat{g}_{\mathbf{k}} = g_{\mathbf{k}}/\sqrt{g_{\mathbf{k}}^2 + \varepsilon_{-,\mathbf{k}}^2}$ and $\hat{\varepsilon}_{-,\mathbf{k}} = \varepsilon_{-,\mathbf{k}}/\sqrt{g_{\mathbf{k}}^2 + \varepsilon_{-,\mathbf{k}}^2}$. In the above expressions, we also used the two (spin-independent) band energies as found in the previous chapter,
$$\xi_{\pm,\mathbf{k}s} = \xi_{\pm,\mathbf{k}} = \varepsilon_{+,\mathbf{k}} - \mu \pm \sqrt{g_{\mathbf{k}}^2 + \varepsilon_{-,\mathbf{k}}^2}. \qquad (4.32)$$

In the following analysis, a pairing interaction of the form
$$\mathcal{H}' = \frac{1}{N}\sum_{\mathbf{k},\mathbf{k}'} V_{\alpha\beta,\mu\nu}^{ss',s_3s_4}(\mathbf{k},\mathbf{k}')c_{\alpha\mathbf{k}s}^{\dagger}c_{\beta-\mathbf{k}s'}^{\dagger}c_{\gamma-\mathbf{k}'s_3}c_{\delta\mathbf{k}'s_4} \qquad (4.33)$$

is used, where the sum over spin and orbital indices is implied again. This interaction can be parametrized in a similar fashion as in the one-band case, Eq. (4.19), leading to
$$V_{\alpha\beta,\mu\nu}^{ss',s_3s_4}(\mathbf{k},\mathbf{k}') = \sum_{m,n}\sum_a v_{mn}^{(a)}[\psi_{mn}^{(a)}(\mathbf{k})\varsigma^m\tau^n]_{\alpha\beta}^{ss'}[\psi_{mn}^{(a)}(\mathbf{k}')\varsigma^m\tau^n]_{\mu\nu}^{\dagger,s_3s_4}, \qquad (4.34)$$

where $\psi_{mn}^{(a)}(\mathbf{k})$ are again parity-allowed basis functions. For a detailed analysis of the structure of such an interaction and a specific example see appendix D. As is discussed there, we have to distinguish the two cases of interactions between quasiparticles sitting on the same sublattice and between quasiparticles on different sublattices and therefore, between intra- and inter-sublattice pairing. We first shortly look at the case of an intra-sublattice interaction on a general level before turning to a nearest-neighbor interaction as an example of an inter-sublattice interaction.

**Intra-sublattice pairing**

For an intra-sublattice interaction, only terms with $\tau^0$ and $\tau^1$ appear in the expansion (4.34), thus allowing for a gap of the form
$$\Delta(\mathbf{k}) = [\psi_0(\mathbf{k})\varsigma^0 + \vec{d}_0(\mathbf{k})\cdot\vec{\varsigma}]\otimes\tau^0 + [\psi_1(\mathbf{k})\varsigma^0 + \vec{d}_1(\mathbf{k})\cdot\vec{\varsigma}]\otimes\tau^1. \qquad (4.35)$$

Due to the symmetry of $\tau^0$ and $\tau^1$, the parity and spin configuration are again related. Looking at tables C.1 and C.2, we thus find that the only coupling appears between the spin-triplet components

$$d_0^x(\mathbf{k}) \leftrightarrow d_1^y(\mathbf{k}) \qquad (4.36)$$
$$d_0^y(\mathbf{k}) \leftrightarrow d_1^x(\mathbf{k}). \qquad (4.37)$$

This could lead to a lifting of the spin-triplet gap degeneracies and thus, fix the direction of the $d$ vector, but as we will see in the following, a calculation with a specific Hamiltonian and interaction is necessary for details.

**Inter-sublattice pairing**

For the case of an inter-sublattice interaction, we find terms $\tau^2$ and $\tau^3$ that are even and odd under interchange of orbital indices, respectively. Therefore, it is necessary to distinguish between an even and an odd part of the gap function with respect to $\mathbf{k}$ instead of spin singlet and triplet, $\Delta(\mathbf{k}) = \Delta_+(\mathbf{k}) + \Delta_-(\mathbf{k})$. Writing the two parts of the gap function as[2]

$$\Delta_-(\mathbf{k}) = \psi_-(\mathbf{k})\varsigma^0 \otimes (i\tau^2) + [\vec{d}_-(\mathbf{k}) \cdot \vec{\varsigma}] \otimes \tau^3 \qquad (4.38)$$

and

$$\Delta_+(\mathbf{k}) = \psi_+(\mathbf{k})\varsigma^0 \otimes \tau^3 + [\vec{d}_+(\mathbf{k}) \cdot \vec{\varsigma}] \otimes (i\tau^2), \qquad (4.39)$$

respectively, we see looking at tables C.1 and C.2 that now the $x$ and $y$ components of the $d$ vector are independent. The corresponding gap equations read

$$d_\pm^{x,y}(\mathbf{k}) = -T \sum_{n,\mathbf{k}'} 4v^\pm(\mathbf{k},\mathbf{k}') d_\pm^{x,y}(\mathbf{k}')(G_{0+}\tilde{G}_{0+} \mp G_{0-}\tilde{G}_{0-}), \qquad (4.40)$$

where we have introduced the short notation $G_{0\pm} = G_{0\pm}(\mathbf{k}',\omega_n)$ and $\tilde{G}_{0\pm} = G_{0\pm}(-\mathbf{k}',-\omega_n)$.

Tables C.1 and C.2 also show that now we have a coupling of the odd $d^z$ component with the odd spin-singlet part and evaluating Eq. (4.27) we obtain

$$\begin{pmatrix} d_-^z(\mathbf{k}) \\ \psi_-(\mathbf{k}) \end{pmatrix} = -T \sum_{n,\mathbf{k}'} 4v^-(\mathbf{k},\mathbf{k}') \begin{pmatrix} M_{11}(\mathbf{k}') & M_{12}(\mathbf{k}') \\ M_{21}(\mathbf{k}') & M_{22}(\mathbf{k}') \end{pmatrix} \begin{pmatrix} d_-^z(\mathbf{k}') \\ \psi_-(\mathbf{k}') \end{pmatrix} \qquad (4.41)$$

---

[2] For a consistent notation, we still use $\psi(\mathbf{k})$ and $\vec{d}(\mathbf{k})$ for the momentum-dependence of the spin-singlet and spin-triplet part of the gap function.

## 4.3 Inversion-Symmetry Lacking Bonds

and similarly for the case of an even gap function

$$\begin{pmatrix} \psi_+(\mathbf{k}) \\ d^z_+(\mathbf{k}) \end{pmatrix} = -T \sum_{n,\mathbf{k}'} 4v^+(\mathbf{k},\mathbf{k}') \begin{pmatrix} M_{11}(\mathbf{k}') & M_{12}(\mathbf{k}') \\ M_{21}(\mathbf{k}') & M_{22}(\mathbf{k}') \end{pmatrix} \begin{pmatrix} \psi_+(\mathbf{k}') \\ d^z_+(\mathbf{k}') \end{pmatrix}. \quad (4.42)$$

The matrix elements are given by

$$M_{11}(\mathbf{k}) = G_{0+}\tilde{G}_{0+} + G_{0-}\tilde{G}_{0-} - 2\hat{g}_\mathbf{k}^2 G_{0-}\tilde{G}_{0-}, \quad (4.43)$$

$$M_{22}(\mathbf{k}) = G_{0+}\tilde{G}_{0+} + G_{0-}\tilde{G}_{0-} - 2\hat{\varepsilon}_{-,\mathbf{k}}^2 G_{0-}\tilde{G}_{0-}, \quad (4.44)$$

$$M_{12}(\mathbf{k}) = 2i\hat{g}_\mathbf{k}\hat{\varepsilon}_{-,\mathbf{k}} G_{0-}\tilde{G}_{0-} = M_{21}^*(\mathbf{k}). \quad (4.45)$$

For further calculations, the following Matsubara sums are useful,

$$S_1(\mathbf{k}) = -T\sum_n (G_{0+}\tilde{G}_{0+} + G_{0-}\tilde{G}_{0-}) = \sum_{a=\pm} \frac{1}{2\xi_{a,\mathbf{k}}} \tanh\left(\frac{\xi_{a,\mathbf{k}}}{2T}\right) \quad (4.46)$$

and

$$S_2(\mathbf{k}) = -T\sum_n (G_{0+}\tilde{G}_{0+} - G_{0-}\tilde{G}_{0-}) = \sum_{a=\pm} \frac{1}{2\varepsilon_{+,\mathbf{k}}} \tanh\left(\frac{\xi_{a,\mathbf{k}}}{2T}\right). \quad (4.47)$$

With these, we can now evaluate the above self-consistency equations explicitly using the nn interaction as derived in appendix D. We first look at the uncoupled $x$ and $y$ component of the $d$ vector and choose $p$-wave pairing for the odd gap, $d^{x,y}_-(\mathbf{k}) = \Delta^{x,y}_- \sin k_x$ (with $\Delta^{x,y}_- \sin k_y$ a degenerate solution) and an extended $s$-wave form for the even gap. From Eq. (4.40) we find using Eq. (4.46) and (4.47),

$$1 = -V \sum_{\mathbf{k}'} \sum_{a=\pm} \frac{\sin^2 k'_x}{2\xi_{a,\mathbf{k}'}} \tanh\left(\frac{\xi_{a,\mathbf{k}'}}{2T}\right), \quad (4.48)$$

$$1 = -V \sum_{\mathbf{k}'} \sum_{a=\pm} \frac{(\cos k'_x + \cos k'_y)^2}{2\varepsilon_{+,\mathbf{k}'}} \tanh\left(\frac{\xi_{a,\mathbf{k}'}}{2T}\right). \quad (4.49)$$

For the coupled gap functions, we start with the odd-parity case and again choose the $p$-wave gap with momentum dependence $\sin k_x$, i.e., $d^z_-(\mathbf{k}) = \Delta^z_- \sin k_x$ and $\psi_-(\mathbf{k}) = \Delta^s_- \sin k_x$. With this choice, we find from Eq. (4.41)

$$\begin{pmatrix} \Delta^z_- \\ \Delta^s_- \end{pmatrix} = \begin{pmatrix} L^-_0 - L^-_1 & iL^-_3 \\ -iL^-_3 & L^-_0 - L^-_2 \end{pmatrix} \begin{pmatrix} \Delta^z_- \\ \Delta^s_- \end{pmatrix}. \quad (4.50)$$

With Eqs. (4.46) and (4.47), we can express these matrix elements as

$$\begin{aligned}
L_0^- &= -V \sum_{\mathbf{k}} \sin^2 k_x S_1(\mathbf{k}), \\
L_1^- &= -V \sum_{\mathbf{k}} \sin^2 k_x \hat{g}_{\mathbf{k}}^2 [S_2(\mathbf{k}) - S_1(\mathbf{k})], \\
L_2^- &= -V \sum_{\mathbf{k}} \sin^2 k_x \hat{\varepsilon}_{-,\mathbf{k}}^2 [S_2(\mathbf{k}) - S_1(\mathbf{k})], \\
L_3^- &= -V \sum_{\mathbf{k}} \sin^2 k_x \hat{g}_{\mathbf{k}} \hat{\varepsilon}_{-,\mathbf{k}} [S_2(\mathbf{k}) - S_1(\mathbf{k})].
\end{aligned} \quad (4.51)$$

### 4.3.4 Discussion

We saw in the above calculation that the appearance of gap couplings depends on the type of the leading pairing interaction. While for an intra-sublattice pairing, the $x$ and $y$ component of the $d$ vector are coupled, for an inter-sublattice interaction the spin-singlet and the $z$ component of the spin-triplet gap are coupled. For a detailed calculation, we chose a nn interaction as it allows for both, even and odd gaps, despite its simplicity.

Due to the logarithmic temperature divergence in Eq. (4.48), there is a finite $T_c$ for any finite and negative interaction strength $V$. Looking at Eq. (4.38) and comparing with table 4.1, we see that this corresponds to an intra-band, inter-sublattice coupling and thus, to a standard zero-momentum gap. On the other hand, the gap equation for $d_+^{x,y}$, Eq. (4.49), has a finite zero-temperature limit therefore requiring a finite strength of the interaction for a non-vanishing transition temperature. This actually corresponds to an inter-band, i.e., finite-momentum gap. The small phase space available for finite-momentum pairing explains the lack of a superconducting instability in this channel.

We now turn to the coupled gap equation (4.50). Naively only looking at the diagonal terms, one would expect as a main consequence of the staggered SOC that both gaps, zero- and finite-momentum, are suppressed. Especially the singlet gap, which is again a finite-momentum gap, shows the same self-consistency condition as the (uncoupled) inter-band gap above, Eq. (4.49). However, for the evaluation of the possible instabilities, the eigenvalue equations following from

Eq. (4.50) should be considered,

$$1 = L_0^- - \frac{1}{2}(L_1^- + L_2^-) \pm \frac{1}{2}\sqrt{(L_1^- - L_2^-)^2 + 4(L_3^-)^2}. \qquad (4.52)$$

Obviously, the higher $T_c$ is obtained for the '+' solution. Using Cauchy-Schwartz inequality, it follows that in general, this solution is smaller or equal to the one obtained before for the $x$ and $y$ components of the $d$ vector. However, the degeneracy is only lifted if $g_\mathbf{k}/\varepsilon_{-,\mathbf{k}}$ depends on the momentum $\mathbf{k}$. For the simple nn- and nnn-hopping Hamiltonian considered in the last chapter we have $g_\mathbf{k}/\varepsilon_{-,\mathbf{k}} = -\alpha/t$ and thus, the gap's spin structure is not fixed. The degeneracy of the $d$-vector components is therefore only lifted in higher orders. The main effect of the staggered SOC is to mix the different spin components, in particular the singlet and the triplet $z$ component, to form new eigenstates of the gap equation.

We can now use the same formalism to study the effect of an antiferromagnetic background interacting with the conduction electrons via a Zeeman term. For this purpose, we replace the SOC term in the Hamiltonian (4.22) by an on-site magnetic term which we choose to point in $z$ direction,

$$\mathcal{H}^{\text{afm}} = H\sigma^z \otimes \tau^1. \qquad (4.53)$$

This term has the same symmetry as the SOC term and indeed going again through the same calculations as above, the same gap couplings are observed. However, for the case of antiferromagnetism, the independent gaps are suppressed. This means that for example for the intra-sublattice pairing, the $x$ and $y$ components of the $d$ vector are coupled and possess an instability. This is not very surprising as these two gaps are also not influenced by a magnetic field in $z$ direction due to their equal-spin coupling in this direction. For the case of inter-sublattice pairing, we find that both, the singlet and the $z$-component spin-triplet gap are almost unaffected by the antiferromagnetism. This is in agreement with the findings of Baltensperger and Straessler [75].

## 4.4 Inversion-Symmetry Lacking Layers

In this second example, we first start with an inversion-symmetry lacking system, a single layer with no mirror symmetry with respect to the plane, and then

*Superconductivity in Crystals with Locally Broken Inversion Symmetry*

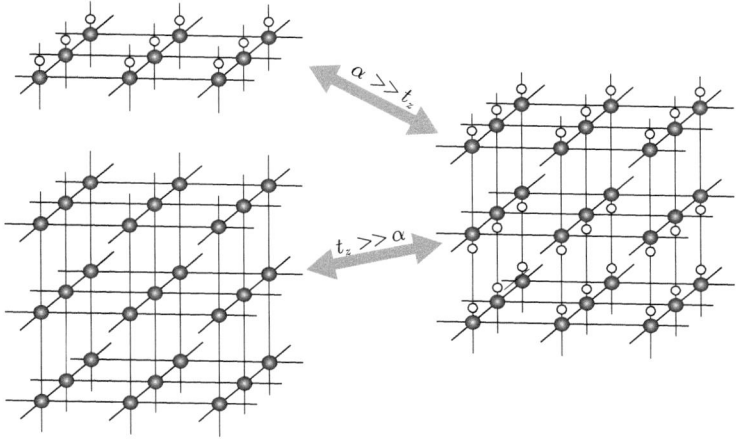

Figure 4.1: Right: The three dimensional crystal consisting of layers with lacking inversion symmetry. On the left side, the two limiting cases are shown, i.e. the single layer without inversion symmetry on top and the three dimensional crystal with $D_{4h}$-symmetry.

arrange these layers in a way as to restore the global inversion symmetry. In the following, we thus first recapitulate briefly the situation of a single-layer system analogue to what was analyzed by Frigeri et al. [82]. Then, we consider the full three-dimensional case and evaluate the linearized gap equation.

### 4.4.1 Single layer

For a (single-layer) square lattice with a broken $\sigma_h$ symmetry, Frigeri et al. derived an additional symmetry-reducing term that appears in the Hamiltonian of

## 4.4 Inversion-Symmetry Lacking Layers

the form
$$\mathcal{H}^{\text{soc}} = \sum_{\mathbf{k}} \sum_{s,s'} (\vec{g}_{\mathbf{k}} \cdot \vec{\sigma}_{ss'}) c^{\dagger}_{\mathbf{k}s} c_{\mathbf{k}s'} \qquad (4.54)$$

with $\vec{g}_{\mathbf{k}} = \alpha(\hat{x}\sin k_y - \hat{y}\sin k_x)$. Similar to the symmetry-broken situation of a sample boundary, this antisymmetric SOC is of Rashba type. Note that opposite to the case analyzed in the preceding section, this term truly breaks inversion symmetry with $\vec{g}_{-\mathbf{k}} = -\vec{g}_{\mathbf{k}}$.

From a symmetry point of view, the lack of mirror symmetry reduces the point group of such a layer from $D_{4h}$ to $C_{4v}$ with the remaining symmetry operations $C_2$, $C_4$, $\sigma_v$ and $\sigma_d$. Therefore, the Hamiltonian can have a symmetry-reducing term belonging to a non-trivial IR in $D_{4h}$ that maps to $A_{1g}$ of $C_{4v}$. Looking at the character table in appendix B, we see that this is $A_{2u}$. The above term (4.54) is indeed of this symmetry. Note that this term is a true SOC term requiring the triplet gap to have a combined momentum and spin part, cf. right column of table B.2.

As in the previous section, the symmetry reduction also leads to the coupling of formerly independent gaps, e.g. the $A_{1g}$ with the $A_{2u}$ gap. This is a singlet - triplet coupling as one would expect from a system lacking inversion symmetry. In order to allow for a direct comparison of the single layer with the full symmetry restored three-dimensional system, appendix E recapitulates the analysis of the linearized gap equation for the single layer. Apart from confirming the above symmetry considerations, the main findings are that (a) the antisymmetric SOC fixes the $d$ vector to the direction of $\vec{g}_{\mathbf{k}}$ and (b) the coupled spin-singlet and spin-triplet gaps both allow for a superconducting instability. This is in contrast to the findings of the previous section, where only the mixed gap had an instability.

### 4.4.2 Symmetry analysis of the stacked planes

We note looking at Fig. 4.1 that choosing the right unit cell (with the center between two layers), the crystal's inversion symmetry is respected and the point group of the crystal is $D_{4h}$. As we are interested in the limiting case of the single layer, we start, however, with a unit cell shifted as to be centered around one of the layers, thus treating the system again as if it was non-symmorphic. Still, the factor group $\mathcal{G}/\mathcal{T}$ is again isomorphic to $D_{4h}$, of course. The advantage of this treatment is that there can actually be terms in the Hamiltonian of the form $A_{2u} \otimes A_{2u}$, where now the Pauli matrices $\tau^1$ and $\tau^2$ for the orbital part belong to the IR $A_{2u}$ while $\tau^0$ and $\tau^3$ still belong to $A_{1g}$. As for the single-layer case, this means that spin-singlet and spin-triplet gaps can couple, e.g.,

$$(A_{2u} \otimes A_{2u}) \leftrightarrow (A_{1g} \otimes A_{1g}), \quad (4.55)$$

$$(A_{2u} \otimes A_{1g}) \leftrightarrow (A_{1g} \otimes A_{2u}). \quad (4.56)$$

However, as the system now has regained inversion symmetry, we would expect the couplings between even and odd gap functions to be strongly suppressed as soon as there is a finite inter-layer hopping.

### 4.4.3 Gap couplings

Following the same steps as in section 4.3, the different superconducting instabilities can be analyzed by looking at the linearized gap equation (4.27). We again cast the Hamiltonian in the form

$$\mathcal{H}^{\text{hop}} = \sum_{\mathbf{k}} \vec{c}_{\mathbf{k}}^{\dagger} \mathcal{H}_{\mathbf{k}}^{\text{hop}} \vec{c}_{\mathbf{k}} \quad (4.57)$$

with $\vec{c}_{\mathbf{k}}^{\dagger} = (c_{\mathbf{k}\uparrow}^{\dagger}, c_{\mathbf{k}\downarrow}^{\dagger}, c_{\mathbf{k}+\mathbf{Q}\uparrow}^{\dagger}, c_{\mathbf{k}+\mathbf{Q}\downarrow}^{\dagger}) \equiv (c_{1\mathbf{k}\uparrow}^{\dagger}, c_{1\mathbf{k}\downarrow}^{\dagger}, c_{2\mathbf{k}\uparrow}^{\dagger}, c_{2\mathbf{k}\downarrow}^{\dagger})$, this time, however, with

$$\mathcal{H}_{\mathbf{k}}^{\text{hop}} = (\varepsilon_{\mathbf{k}}^{xy} - \mu)\sigma^0 \otimes \tau^0 + \varepsilon_{\mathbf{k}}^{z}\sigma^0 \otimes \tau^3 + \vec{g}_{\mathbf{k}} \cdot \vec{\sigma} \otimes \tau^1 \quad (4.58)$$

and $\mathbf{Q} = (0, 0, \pi)$. Here, $\varepsilon_{\mathbf{k}}^{xy} = -2t_{xy}(\cos k_x + \cos k_y) - 4t'_{xy}\cos k_x \cos k_y$ is the intra-layer-hopping energy and $\varepsilon_{\mathbf{k}}^z = -2t_z \cos k_z$ is the interlayer-hopping energy. The spectrum of the Hamiltonian is thus given by

$$\xi_{\pm,\mathbf{k}} = \varepsilon_{\mathbf{k}}^{xy} - \mu \pm \sqrt{(\varepsilon_{\mathbf{k}}^z)^2 + |\vec{g}_{\mathbf{k}}|^2}. \quad (4.59)$$

The non-interacting Green's function has also almost the same form

$$G_0(\mathbf{k}, \omega_n) = G_{0+}(\mathbf{k}, \omega_n)\sigma^0 \otimes \tau^0 + G_{0-}(\mathbf{k}, \omega_n)(\hat{g}_{\mathbf{k}} \cdot \vec{\sigma} \otimes \tau^1 + \hat{\varepsilon}_{\mathbf{k}}^z \sigma^0 \otimes \tau^3) \quad (4.60)$$

where

$$G_{0\pm}(\mathbf{k}, \omega_n) = \frac{1}{2}\left(\frac{1}{i\omega_n - \xi_{+,\mathbf{k}}} \pm \frac{1}{i\omega_n - \xi_{-,\mathbf{k}}}\right) \quad (4.61)$$

## 4.4 Inversion-Symmetry Lacking Layers

and we have defined $\hat{g}_{\mathbf{k}} = \vec{g}_{\mathbf{k}}/\sqrt{(\varepsilon_{\mathbf{k}}^z)^2 + |\vec{g}_{\mathbf{k}}|^2}$ and $\hat{\varepsilon}_{\mathbf{k}}^z = \varepsilon_{\mathbf{k}}^z/\sqrt{(\varepsilon_{\mathbf{k}}^z)^2 + |\vec{g}_{\mathbf{k}}|^2}$.

As we saw in the previous section, we need to distinguish the two cases of an intra-sublattice (here intra-layer) and inter-sublattice pairing interaction. For this kind of a system we expect a rather drastic difference for the two cases, as one could imagine that for intra-layer pairing and sufficiently small inter-layer coupling, the system still behaves like a system of disconnected single layers with a lacking inversion symmetry.

### Intra-layer interaction

As in the previous section, the intra-sublattice interaction allows for a gap of the form

$$\Delta(\mathbf{k}) = [\psi_0(\mathbf{k})\varsigma^0 + \vec{d}_0(\mathbf{k}) \cdot \vec{\varsigma}] \otimes \tau^0 + [\psi_1(\mathbf{k})\varsigma^0 + \vec{d}_1(\mathbf{k}) \cdot \vec{\varsigma}] \otimes \tau^1. \tag{4.62}$$

Similar to the case of a single mirror-symmetry-lacking layer, a coupling of the spin-singlet and spin-triplet gaps is found,

$$\psi_0(\mathbf{k}) = -T \sum_{n,\mathbf{k}'} v^+(\mathbf{k},\mathbf{k}') \Big\{ [G_{0+}\tilde{G}_{0+} + G_{0-}\tilde{G}_{0-}]\psi_0(\mathbf{k}')$$
$$+ [G_{0+}\tilde{G}_{0-} + G_{0-}\tilde{G}_{0+}]\hat{g}_{\mathbf{k}'} \cdot \vec{d}_1(\mathbf{k}') \Big\}, \tag{4.63}$$

$$\vec{d}_1(\mathbf{k}) = -T \sum_{n,\mathbf{k}'} v^-(\mathbf{k},\mathbf{k}') \Big\{ [G_{0+}\tilde{G}_{0+} - G_{0-}\tilde{G}_{0-}]\vec{d}_1(\mathbf{k}')$$
$$+ 2G_{0-}\tilde{G}_{0-}[\hat{g}_{\mathbf{k}'} \cdot \vec{d}_-(\mathbf{k}')]\hat{g}_{\mathbf{k}'} + [G_{0+}\tilde{G}_{0-} + G_{0-}\tilde{G}_{0+}]\vec{g}_{\mathbf{k}'}\psi_0(\mathbf{k}') \Big\} \tag{4.64}$$

and also

$$\vec{d}_0(\mathbf{k}) = -T \sum_{n,\mathbf{k}'} v^-(\mathbf{k},\mathbf{k}') \Big\{ [G_{0+}\tilde{G}_{0+} + G_{0-}\tilde{G}_{0-}]\vec{d}_0(\mathbf{k}')$$
$$+ 2G_{0-}\tilde{G}_{0-}\{\hat{g}_{\mathbf{k}'}[\hat{g}_{\mathbf{k}'} \cdot \vec{d}(\mathbf{k}')] - \vec{d}(\mathbf{k}')\} + [G_{0+}\tilde{G}_{0-} + G_{0-}\tilde{G}_{0+}]\vec{g}_{\mathbf{k}'}\psi_1(\mathbf{k}') \Big\}, \tag{4.65}$$

$$\psi_1(\mathbf{k}) = -T \sum_{n,\mathbf{k}'} v^+(\mathbf{k},\mathbf{k}') \Big\{ [G_{0+}\tilde{G}_{0+} + G_{0-}\tilde{G}_{0-} - 2(\hat{\varepsilon}_{\mathbf{k}'}^z)^2 G_{0-}\tilde{G}_{0-}]\psi_1(\mathbf{k}')$$
$$+ [G_{0+}\tilde{G}_{0-} + G_{0-}\tilde{G}_{0+}]\hat{g}_{\mathbf{k}'} \cdot \vec{d}_0(\mathbf{k}') \Big\}. \tag{4.66}$$

**Interlayer interaction**

The terms $\tau^2$ and $\tau^3$ appearing in the interlayer interaction are even and odd under interchange of orbital indices, respectively. Therefore, a distinction in even and odd gap functions in $\mathbf{k}$ instead of spin singlet and spin triplet is adopted again, $\Delta(\mathbf{k}) = \Delta_+(\mathbf{k}) + \Delta_-(\mathbf{k})$. Continuing as before, the two parts of the gap function are written as

$$\Delta_-(\mathbf{k}) = \psi_-(\mathbf{k})\varsigma^0 \otimes (i\tau^2) + [\vec{d}_-(\mathbf{k}) \cdot \vec{\varsigma}] \otimes \tau^3 \tag{4.67}$$

and

$$\Delta_+(\mathbf{k}) = \psi_+(\mathbf{k})\varsigma^0 \otimes \tau^3 + [\vec{d}_+(\mathbf{k}) \cdot \vec{\varsigma}] \otimes (i\tau^2). \tag{4.68}$$

However, we note that the two parts $\psi_-(\mathbf{k})\sigma^0 \otimes (i\tau^2)$ and $[\vec{d}_+(\mathbf{k}) \cdot \vec{\varsigma}] \otimes (i\tau^2)$, where the parity of the momentum dependent part is required by the Pauli principle, are neither $A_{1g}$ nor $A_{2u}$. Therefore, there is no symmetry-allowed coupling anymore and the spin-singlet and spin-triplet part are completely decoupled. The respective linearized gap equations for the inter-band pairing read

$$\psi_+(\mathbf{k}) = -T \sum_{n,\mathbf{k}'} v^+(\mathbf{k}, \mathbf{k}')[G_{0+}\tilde{G}_{0+} + (\hat{\varepsilon}_{\mathbf{k}'}^2 - \hat{g}_{\mathbf{k}'}^2)G_{0-}\tilde{G}_{0-}]\psi_+(\mathbf{k}') \tag{4.69}$$

and also

$$\vec{d}_-(\mathbf{k}) = -T \sum_{n,\mathbf{k}'} v^-(\mathbf{k}, \mathbf{k}') \Big\{ [G_{0+}\tilde{G}_{0+} + G_{0-}\tilde{G}_{0-}]\vec{d}(\mathbf{k}')$$
$$- 2G_{0-}\tilde{G}_{0-}[\hat{g}_{\mathbf{k}'} \cdot \vec{d}_-(\mathbf{k}')]\hat{g}_{\mathbf{k}'} \Big\}. \tag{4.70}$$

### 4.4.4 Discussion

Focussing on the diagonal terms in Eqs. (4.63)-(4.66) it is clear that as in the single-layer case, the gaps with a $d$ vector not parallel to $\vec{g}_\mathbf{k}$ are suppressed for an intra-layer pairing interaction. However, evaluating the intra-band gap equations for a nearest-neighbor intra-layer pairing interaction, our results show that $T_c$ is less suppressed for finite inter-layer hopping. This is shown in Fig. 4.2. The completely decoupled layers, $t_z = 0$, corresponds to the case of a system globally lacking inversion symmetry. When looking at the coupled intra-band gaps, we

## 4.4 Inversion-Symmetry Lacking Layers

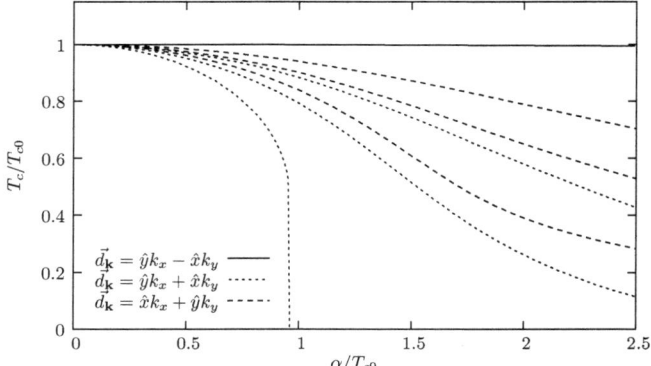

Figure 4.2: Suppression of the transition temperature of the intra-band gaps due to the antisymmetric SOC. With increasing inter-layer coupling, the suppression is weakened, $t_z = 0, 0.1t, 0.2t$ from bottom to top.

see that opposite to the single-layer case, one of the gaps in each case [$\psi_1(\mathbf{k})$ and $\vec{d}_1(\mathbf{k})$, respectively] is now suppressed as it involves finite-momentum pairing. How this decreases the critical temperature $T_c$ is shown in Fig. 4.3, where also the effect of weakening this suppression by the antisymmetric SOC is revealed.

Eventually, we arrive at a picture, where with increasing interlayer coupling the superconducting instability moves further away from the singlet-triplet coupling known from the inversion-symmetry-lacking system towards a behavior reminiscent of a globally inversion-invariant crystal with $D_{4h}$ symmetry. Finally, when the coupling between the layers becomes strong enough for the leading instability to be in the interlayer channel, the system retains its full symmetry. Interestingly, in this case, the antisymmetric SOC suppresses the spin-singlet gap and the favored gap is the Balian-Werthamer gap with $\vec{d}_\mathbf{k} = \hat{x} k_x + \hat{y} k_y + \hat{z} k_z$.

Also, the effect of an alternating magnetization in the planes can be analyzed by changing the SOC term to a staggered magnetization term,

$$\mathcal{H}^{\text{afm}} = \vec{H} \cdot \vec{\sigma} \otimes \tau^1. \tag{4.71}$$

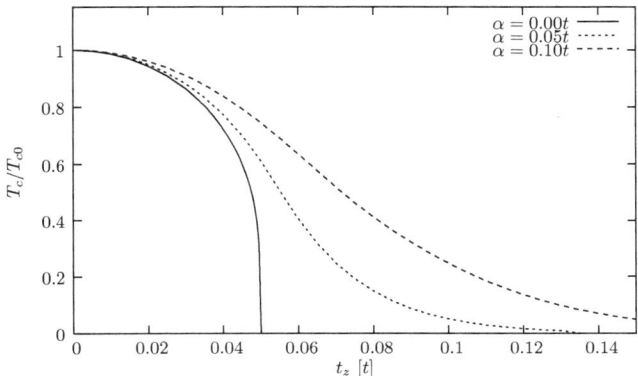

Figure 4.3: Change in the transition temperature of the inter-band gap with $\vec{d}_1(\mathbf{k}) = \hat{x}\sin k_y - \hat{y}\sin k_x$ as a function of the inter-layer hopping $t_z$ for different spin-orbit coupling strengths.

For the case of an intra-layer coupling interaction, the intra-band spin-triplet gaps with a $d$ vector perpendicular to $\vec{H}$ are unaffected and couple to the corresponding inter-band gaps. On the other hand, the spin-singlet gap is again hardly affected for the inter-layer coupling and it mixes with the inter-band triplet gap with $\vec{d}(\mathbf{k}) \parallel \vec{H}$ and vice versa.

## 4.5 Conclusions and Outlook

In this chapter, a detailed analysis of superconductivity in crystals with locally broken inversion symmetry, but global inversion invariance was presented. The global invariance arises due to the periodic arrangement of the symmetry breaking elements and resembles therefore the case of superconductivity in antiferromagnetic materials. The two-band formulation used is therefore easily transferable to this case and is similar to the one used in the literature (see e.g. Refs. [75, 84, 85, 86]).

## 4.5 Conclusions and Outlook

The two cases studied in sections 4.3 and 4.4 are in spite of some similarities quite different. As in both examples the original symmetry is reduced, formerly independent gap functions are now coupled. However, while in the first example spin-singlet and spin-triplet gaps can mix forming gaps that can still be separated into even and odd momentum dependent functions, in the second example, even and odd gap functions mix themselves. In addition, while the coupling is important for the evaluation of the instability conditions for superconductivity in the first case, it only depends on the particle-hole asymmetry in the second case and can be neglected to first order. The picture for this latter case is therefore rather a singlet (triplet) zero-momentum gap inducing a triplet (singlet) finite-momentum gap while for the former case, the mixed state itself has to be considered for exploring the instability. A further interesting aspect is the lifting of degeneracies and thus, the fixation of a direction of the $d$ vector. The two cases studied are very different even though the origin of the symmetry reducing terms is for both the relativistic on-site spin-orbit coupling (so-called $L$-$S$ coupling). While the effect of the SOC in the first example was only minor, the antisymmetric SOC in the second example suppresses all but one direction of the $d$ vector, as was seen by studying the diagonal part of the linearized gap equation. This is due to the fact that the antisymmetric SOC splits the Fermi surfaces even for the case of no external magnetic field. Lastly, while the coupling of the symmetry-broken elements in the first case is intrinsic - one would in general expect the inter-sublattice (nn) hopping to be at least comparable to the intra-sublattice (nnn) hopping - the second case allows for the study of the crossover from a global symmetry lacking system to one which is globally inversion-symmetric through the ratio $\alpha/t_z$. This results as well as the comparison to antiferromagnetic superconductors are summarized in table 4.2.

For future studies, the examination of the effect of antiferro-type ordered symmetry reduction on the upper critical field $H_{c2}$ seems also very promising. In general, an applied magnetic field suppresses the transition temperature for the spin-singlet gap and the spin-triplet gap with $d$ vector in the field direction. In the second example, it should simply become easier to rotate the $d$ vector with an applied field, since the $d$-vector direction is fixed less rigid for finite inter-layer coupling. The first example, however, appears to be more interesting: Not only is the gap function to be considered a mix of the $z$-component spin-triplet gap

*Superconductivity in Crystals with Locally Broken Inversion Symmetry*

|  | symmetry broken | | |
|---|---|---|---|
|  | time-reversal | inversion | |
| ferro-type | $\vec{d} \perp \vec{H}$ | $\psi \leftrightarrow (\vec{d} \parallel \vec{g})$ | |
|  | | bonds | layers |
| antiferro-type | $(\vec{d}_{0/1} \perp \vec{H}) \rightleftarrows (\vec{d}_{1/0} \perp \vec{H})$ $\psi_\pm \rightleftarrows (\vec{d}_\pm \parallel \vec{H})$ | $\psi_0$, $\vec{d}_0 \parallel \vec{g}$ $\vec{d}_+ \perp \vec{g}$ $\psi_\pm \leftrightsquigarrow (\vec{d}_\pm \parallel \vec{g})$ $(\vec{d}_{0/1} \perp \vec{H}) \leftrightsquigarrow (\vec{d}_{1/0} \perp \vec{g})$ | $\psi_{0/1} \rightleftarrows (\vec{d}_{1/0} \parallel \vec{g})$ $\vec{d}_+ \perp \vec{g}$ |

Table 4.2: Comparison of allowed gap functions in systems lacking time-reversal and inversion symmetry, respectively. The notation for the different gap functions follows the one adopted in this chapter, e.g., $\psi_1(\mathbf{k})$ is the intra-sublattice inter-band spin-singlet gap function. In addition, the single arrow denotes a mixing of equal gaps and two arrows pointing in opposite directions indicate one of the gaps inducing the other. Finally, the wiggled arrow denotes the case of gaps only possessing an instability as a mixed new eigenfunction. These last gaps are actually suppressed, but only in higher order.

with the spin-singlet gap, but also opens the induced SDW gaps in the Fermi surface, thus additionally suppressing the superconducting instability.

## 4.5 Conclusions and Outlook

# Appendix A

# SDW Instability

As was discussed in the main text, the proximity of the system to a SDW instability is crucial for obtaining a sizeable anisotropy in the response to a magnetic field. However, when the system is too close to a SDW instability, it is possible to drive it through this instability even when a field is applied in $z$ direction. This is, however, not the aim of this work. In this appendix we therefore want to analyze the parameter region where we can avoid the occurrence of an in-plane SDW for this case before reaching the metamagnetic transition. To be more specific, the proximity to a SDW instability is analyzed as a function of density for fixed nnn hopping $t' = 0.36t$.

For this purpose, we first need to include the possibility of a staggered magnetization in plane in the Hamiltonian with a $z$-direction magnetic field. Since a staggered magnetic moment in $y$ direction couples to a homogeneous magnetic field in $x$ direction through the staggered SOC, we also have to include a term for such a magnetization. We start from the Hubbard-$U$ term in Eq. (3.23), but now consider spin-quantization axes that are staggered in $y$-direction, while homogeneous in $x$ and $z$ direction,

$$\begin{aligned}
\mathcal{H}^U &= -U \sum_{i \in A} S_i^{\hat{a}} S_i^{\hat{a}} - U \sum_{i \in B} S_i^{\hat{b}} S_i^{\hat{b}} \\
&= UNM^2 - 2UM \sum_{i \in A} (S_i^x \sin\theta \cos\phi + S_i^y \sin\theta \sin\phi + S_i^z \cos\theta) \\
&\quad - 2UM \sum_{i \in B} (S_i^x \sin\theta \cos\phi - S_i^y \sin\theta \sin\phi + S_i^z \cos\theta). \quad \text{(A.1)}
\end{aligned}$$

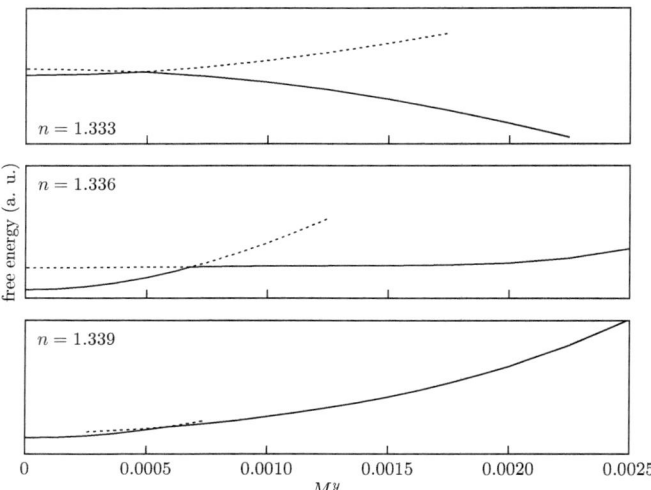

Figure A.1: The free energy for an on-site interaction term as given in Eq. (A.1) as a function of $M^y$ for the three densities $n = 1.333$, $n = 1.336$ and $n = 1.339$. The solid line denotes the thermodynamic (minimal) free energy while the dashed lines correspond to metastable solutions. This metastable solutions appear due to the proximity of the system to a first-order metamagnetic transition. Obviously, the free energy for the densities $n = 1.336$ and $n = 1.339$ is minimal for the case of no staggered magnetization. However, for the case of $n = 1.333$, the free energy could be minimized by a staggered magnetization. The system would thus be driven through a SDW instability.

## SDW Instability

To check for the possible occurrence of a SDW, we fix the magnetic field at a value just below the critical strength $H_c^z$ and evaluate the free energy per lattice site $f_{M^x M^z}(M^y)$ as a function of the staggered magnetic moment with the homogeneous magnetization in $x$ and $z$ direction calculated self consistently. Note that, similar to the main text, we chose the order parameters $M^x$, $M^y$ and $M^z$ instead of $M$, $\phi$ and $\theta$. For the numerical analysis, the three densities $n = 1.339$, $n = 1.336$ and $n = 1.333$ with respective on-site interaction strength $U$ with $U_c^{FM} - U = \text{const}$ are chosen, as is indicated for the former two densities in chapter 3 Fig. 3.6. Therefore, a smaller density implies being closer to a SDW instability. The results of this calculation are shown in Fig. A.1. Due to the proximity to a first-order metamagnetic transition, the self-consistency equations have two solutions. The lower one of the free energies obtained from these solutions is the true thermodynamic potential (solid lines). The free energy of the metastable solution is also shown (dashed lines) for completeness. For the values of $n$ chosen in chapter 3, the minimum of the free energy is at $M^y = 0$ and there is no SDW instability occurring before the metamagnetic transition. For the density $n = 1.333$, however, the system indeed has an instability towards the formation of a SDW before the critical magnetic field is reached as Fig. A.1 shows.

# Appendix B

# Character Tables for $D_{4h}$

For both case studies in chapter 4, the original crystal structure is $D_{4h}$. Therefore, the character table for this point group as well as a list of basis functions for the case of no SOC and strong SOC are given in this Appendix. In addition, Fig. B.1(a) shows the stereographic projections of $D_{4h}$ with its four-fold rotation axis around $z$, the four two-fold rotation axes and the mirror planes $\sigma_v$, $\sigma_d$ and $\sigma_h$ ($\oplus$ indicates the coincidence of the projection of points above and below the plane). Also shown is in Fig. B.1(b) the stereographic projection for $C_{4v}$, relevant for the discussion in section 4.4. This point group has a four-fold rotation axis around $z$ and two mirror planes $\sigma_v$ and $\sigma_d$.

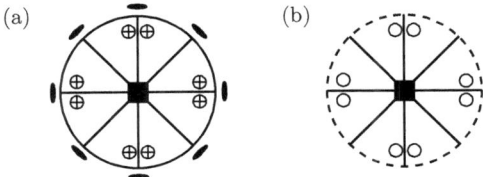

Figure B.1: Stereographic projections for the two point groups $D_{4h}$ (a) and $C_{4v}$ (b).

| $D_{4h}$ | E | $C_2$ | $2C_4$ | $2C_2'$ | $2C_2''$ | I | $\sigma_h$ | $2S_4$ | $2\sigma_v$ | $2\sigma_d$ |
|---|---|---|---|---|---|---|---|---|---|---|
| $A_{1g}$ | 1 | 1 | 1 | 1 | 1 | 1 | 1 | 1 | 1 | 1 |
| $A_{2g}$ | 1 | 1 | 1 | -1 | -1 | 1 | 1 | 1 | -1 | -1 |
| $B_{1g}$ | 1 | 1 | -1 | 1 | -1 | 1 | 1 | -1 | 1 | -1 |
| $B_{2g}$ | 1 | 1 | -1 | -1 | 1 | 1 | 1 | -1 | -1 | 1 |
| $E_g$ | 2 | -2 | 0 | 0 | 0 | 2 | -2 | 0 | 0 | 0 |
| $A_{1u}$ | 1 | 1 | 1 | 1 | 1 | -1 | -1 | -1 | -1 | -1 |
| $A_{2u}$ | 1 | 1 | 1 | -1 | -1 | -1 | -1 | -1 | 1 | 1 |
| $B_{1u}$ | 1 | 1 | -1 | 1 | 1 | -1 | -1 | 1 | -1 | 1 |
| $B_{2u}$ | 1 | 1 | -1 | -1 | -1 | -1 | -1 | 1 | 1 | -1 |
| $E_u$ | 2 | -2 | 0 | 0 | 0 | -2 | 2 | 0 | 0 | 0 |

Table B.1: Character table for the point group $D_{4h}$. This is the symmetry group of the two examples in chapter 4.

| $D_{4h}$ | no SOC | with SOC |
|---|---|---|
| $A_{1g}$ | $\psi = 1$ | $\psi = 1$ |
| $A_{2g}$ | $\psi = k_x k_y (k_x^2 - k_y^2)$ | $\psi = k_x k_y (k_x^2 - k_y^2)$ |
| $B_{1g}$ | $\psi = k_x^2 - k_y^2$ | $\psi = k_x^2 - k_y^2$ |
| $B_{2g}$ | $\psi = k_x k_y$ | $\psi = k_x k_y$ |
| $E_g$ | $\psi = \{k_x k_z, k_y k_z\}$ | $\psi = \{k_x k_z, k_y k_z\}$ |
| $A_{1u}$ | $d^i = (x^2 - y^2)xyz$ | $\vec{d} = \hat{x} k_x + \hat{y} k_y$ |
| $A_{2u}$ | $d^i = k_z$ | $\vec{d} = \hat{x} k_y - \hat{y} k_x$ |
| $B_{1u}$ | $d^i = k_x k_y k_z$ | $\vec{d} = \hat{x} k_x - \hat{y} k_y$ |
| $B_{2u}$ | $d^i = (k_x^2 - k_y^2) k_z$ | $\vec{d} = \hat{x} k_y + \hat{y} k_x$ |
| $E_u$ | $d^i = \{k_x, k_y\}$ | $\vec{d} = \{\hat{z} k_x, \hat{z} k_y\}$ |

Table B.2: The lowest-order basis functions for the gaps of a system where rotations in spin and momentum space are unrelated (left column) and a system where they are coupled (right column). Note that the components for the $d$ vector on the left are independent and we thus have three degenerate solutions for every IR.

# Appendix C

# Coupling of Different Gap Functions

When parametrizing the gap function in terms of Pauli matrices for spin and orbital degrees of freedom as in Eq. (4.23), we derived the allowed gap couplings in section 4.3 by means of a symmetry analysis, see Eqs. (4.24)-(4.26). This means that for example $\Delta_{y0}(\mathbf{k})$ [belonging to $(E_g \times A_{1g})$] couples to $\Delta_{x1}(\mathbf{k})$ [belonging to $(E_g \times A_{2g})$]. Additionally parametrizing the pairing interaction and the non-interaction Green's function in the same way and using the linearized gap equation (4.27), this symmetry analysis yields basically the same result as the condition

$$\sum_{\alpha\beta}\sum_{ss'}\left\{\sum_{kl}\sum_{op} G_0^{kl} G_0^{op} \left[(\sigma^k \otimes \tau^l)(\varsigma^m \otimes \tau^n)(\sigma^o \otimes \tau^p)\right]_{\alpha\beta}^{ss'} (\varsigma^i \otimes \tau^j)_{\alpha\beta}^{ss'}\right\} \neq 0$$
(C.1)

for an allowed coupling. Since $G_0(\mathbf{k})$ is even under $\mathbf{k} \mapsto -\mathbf{k}$, we find the additional condition that $\Delta_{ij}(\mathbf{k})$ and $\Delta_{mn}(\mathbf{k})$ can only couple when their parity with respect to $\mathbf{k}$ is equal. For the Green's function as given in Eq. (4.30), the possible couplings are summarized in table C.1 for odd gap functions and C.2 for even gap functions. Note that the gap functions are further subdivided into the ones supported by an intra-sublattice interaction (top-left) and the ones supported by an inter-sublattice interaction (bottom-right).

| $\Delta_{ij}(\mathbf{k})$ | (y,0) | (z,0) | (x,0) | (y,1) | (z,1) | (x,1) | (y,3) | (z,3) | (x,3) | (0,2) |
|---|---|---|---|---|---|---|---|---|---|---|
| (y,0) | × | - | - | - | - | × | × | - | - | - |
| (z,0) | - | × | - | - | - | - | - | × | - | - |
| (x,0) | - | - | × | × | - | - | - | - | × | - |
| (y,1) | - | - | × | × | - | - | - | - | × | - |
| (z,1) | - | - | - | - | × | - | - | - | - | - |
| (x,1) | × | - | - | - | - | × | × | - | - | - |
| (y,3) | × | - | - | - | - | × | × | - | - | - |
| (z,3) | - | × | - | - | - | - | - | × | - | × |
| (x,3) | - | - | × | × | - | - | - | - | × | - |
| (0,2) | - | × | - | - | - | - | - | × | - | × |

Table C.1: Possible couplings of gap functions, where $(i,j)$ denotes the gap function $\Delta_{ij}(\mathbf{k})$ for the situation described in section 4.3 with the crosses denoting allowed couplings. Only gaps with an odd parity with respect to $\mathbf{k}$ are listed here.

| $\Delta_{ij}(\mathbf{k})$ | (0,0) | (0,1) | (y,2) | (z,2) | (x,2) | (0, 3) |
|---|---|---|---|---|---|---|
| (0,0) | × | - | - | × | - | × |
| (0,1) | - | × | - | - | - | - |
| (y,2) | - | - | × | - | - | - |
| (z,2) | × | - | - | × | - | × |
| (x,2) | - | - | - | - | × | - |
| (0,3) | × | - | - | × | - | × |

Table C.2: Possible couplings of gap functions, where $(i,j)$ denotes the gap function $\Delta_{ij}(\mathbf{k})$ for the situation described in section 4.3 with the crosses denoting allowed couplings again. Only gaps with an even parity with respect to $\mathbf{k}$ are listed here.

# Appendix D

# Structure of the Pairing Interaction

In this appendix, the structure of a general density-density interaction in a system with a two-site unit cell is analyzed. The generalization to other types of interactions, e.g., a spin-spin interaction, is straightforward. We start from a real-space formulation of the interaction,

$$\mathcal{H}' = \sum_{i,j}\sum_{s,s'} V_{ij} n_{is} n_{js'} = \sum_{i,j}\sum_{s,s'} V_{ij} c^\dagger_{is} c^\dagger_{js'} c_{js'} c_{is} \tag{D.1}$$

with $V_{ij}$ the interaction strength between the lattice sites $i$ and $j$. Note that for the special case of $i = j$ (on-site interaction) the spin sum only runs over $s \neq s'$.

Changing now to momentum space, Eq. (D.1) yields

$$\mathcal{H}' = \frac{1}{N} \sum_{\mathbf{k},\mathbf{k}',\mathbf{q}}\sum_{s,s'} v(\mathbf{k},\mathbf{k}') c^\dagger_{\mathbf{k}s} c^\dagger_{-\mathbf{k}+\mathbf{q}s'} c_{-\mathbf{k}'+\mathbf{q}s'} c_{\mathbf{k}'s}, \tag{D.2}$$

where $v(\mathbf{k},\mathbf{k}') = v(\mathbf{k}-\mathbf{k}')$ due to translational symmetry of the crystal. Since we are interested in a situation with two sites per unit cell, we introduce two species of electron operators,

$$c^\dagger_{\alpha\mathbf{k}s} = \begin{cases} c^\dagger_{\mathbf{k}s} & \alpha = 1, \\ c^\dagger_{\mathbf{k}+\mathbf{Q}s} & \alpha = 2, \end{cases} \tag{D.3}$$

where $\mathbf{Q} = (\pi,\pi)$ for a system as described in section 4.3 and $\mathbf{Q} = (0,0,\pi)$ for the situation of section 4.4, respectively. Accordingly, we restrict the sum in

Eq. (D.2) to the two cases of $\mathbf{q} = 0$ and $\mathbf{q} = \mathbf{Q}$ in the following. For the case $\mathbf{q} = 0$ we find

$$\mathcal{H}'_0 = \frac{1}{N} \sum_{\mathbf{kk'}} \{v(\mathbf{k}-\mathbf{k'})[c^\dagger_{1\mathbf{k}s} c^\dagger_{1-\mathbf{k}s'} c_{1-\mathbf{k'}s'} c_{1\mathbf{k'}s} + c^\dagger_{2\mathbf{k}s} c^\dagger_{2-\mathbf{k}s'} c_{2-\mathbf{k'}s'} c_{2\mathbf{k'}s}]$$
$$+ v(\mathbf{k}-\mathbf{k'}+\mathbf{Q})[c^\dagger_{1\mathbf{k}s} c^\dagger_{1-\mathbf{k}s'} c_{2-\mathbf{k'}s'} c_{2\mathbf{k'}s} + c^\dagger_{2\mathbf{k}s} c^\dagger_{2-\mathbf{k}s'} c_{1-\mathbf{k'}s'} c_{1\mathbf{k'}s}]\}. \quad (\text{D.4})$$

For the other case $\mathbf{q} = \mathbf{Q}$, the interaction term can similarly be written as

$$\mathcal{H}'_\mathbf{Q} = \frac{1}{N} \sum_{\mathbf{kk'}} \{v(\mathbf{k}-\mathbf{k'})[c^\dagger_{1\mathbf{k}s} c^\dagger_{2-\mathbf{k}s'} c_{2-\mathbf{k'}s'} c_{1\mathbf{k'}s} + c^\dagger_{2\mathbf{k}s} c^\dagger_{1-\mathbf{k}s'} c_{1-\mathbf{k'}s'} c_{2\mathbf{k'}s}]$$
$$+ v(\mathbf{k}-\mathbf{k'}+\mathbf{Q})[c^\dagger_{1\mathbf{k}s} c^\dagger_{2-\mathbf{k}s'} c_{1-\mathbf{k'}s'} c_{2\mathbf{k'}s} + c^\dagger_{2\mathbf{k}s} c^\dagger_{1-\mathbf{k}s'} c_{2-\mathbf{k'}s'} c_{1\mathbf{k'}s}]\}. \quad (\text{D.5})$$

At this point, a distinction into two cases has to be made:

(a) For an interaction between sites belonging to the same sublattice, $i, j \in \mathcal{A}$ ($\mathcal{B}$), we have $v(\mathbf{k}+\mathbf{Q}) = v(\mathbf{k})$ and the above expressions can be written as

$$\mathcal{H}'_0 = \frac{1}{N} \sum_{\alpha\beta\gamma\delta} \sum_{\mathbf{k},\mathbf{k'}} v(\mathbf{k}-\mathbf{k'})[(\tau^0)_{\alpha\beta}(\tau^0)^\dagger_{\gamma\delta}] c^\dagger_{\alpha\mathbf{k}s} c^\dagger_{\beta-\mathbf{k}s'} c_{\gamma-\mathbf{k'}s'} c_{\delta\mathbf{k'}s}, \quad (\text{D.6})$$

$$\mathcal{H}'_\mathbf{Q} = \frac{1}{N} \sum_{\alpha\beta\gamma\delta} \sum_{\mathbf{k},\mathbf{k'}} v(\mathbf{k}-\mathbf{k'})[(\tau^1)_{\alpha\beta}(\tau^1)^\dagger_{\gamma\delta}] c^\dagger_{\alpha\mathbf{k}s} c^\dagger_{\beta-\mathbf{k}s'} c_{\gamma-\mathbf{k'}s'} c_{\delta\mathbf{k'}s}. \quad (\text{D.7})$$

(b) For an interaction between sites belonging to different sublattices, $i \in \mathcal{A}(\mathcal{B})$ and $j \in \mathcal{B}(\mathcal{A})$, $v(\mathbf{k}+\mathbf{Q}) = -v(\mathbf{k})$ and the above expressions read

$$\mathcal{H}'_0 = \frac{1}{N} \sum_{\alpha\beta\gamma\delta} \sum_{\mathbf{k},\mathbf{k'}} v(\mathbf{k}-\mathbf{k'})[(\tau^3)_{\alpha\beta}(\tau^3)^\dagger_{\gamma\delta}] c^\dagger_{\alpha\mathbf{k}s} c^\dagger_{\beta-\mathbf{k}s'} c_{\gamma-\mathbf{k'}s'} c_{\delta\mathbf{k'}s}, \quad (\text{D.8})$$

$$\mathcal{H}'_\mathbf{Q} = \frac{1}{N} \sum_{\alpha\beta\gamma\delta} \sum_{\mathbf{k},\mathbf{k'}} v(\mathbf{k}-\mathbf{k'})[(i\tau^2)_{\alpha\beta}(i\tau^2)^\dagger_{\gamma\delta}] c^\dagger_{\alpha\mathbf{k}s} c^\dagger_{\beta-\mathbf{k}s'} c_{\gamma-\mathbf{k'}s'} c_{\delta\mathbf{k'}s}. \quad (\text{D.9})$$

In addition, also Pauli matrices for the spin degrees of freedom can be introduced to find for the respective sums

$$\sum_{ss'} c^\dagger_{\alpha\mathbf{k}s} c^\dagger_{\beta-\mathbf{k}s'} c_{\gamma-\mathbf{k'}s'} c_{\delta\mathbf{k'}s} = \frac{1}{2} \sum_{s_1...s_4} \Lambda_{s_1 s_2 s_3 s_4} c^\dagger_{\alpha\mathbf{k}s_1} c^\dagger_{\beta-\mathbf{k}s_2} c_{\gamma-\mathbf{k'}s_3} c_{\delta\mathbf{k'}s_4}, \quad (\text{D.10})$$

where
$$\Lambda_{s_1s_2s_3s_4} = (\varsigma^0)_{s_1s_2}(\varsigma^0)^\dagger_{s_3s_4} + (\vec{\varsigma})_{s_1s_2} \cdot (\vec{\varsigma})^\dagger_{s_3s_4} \tag{D.11}$$
consists of a singlet and a triplet part considering the exchange of spin indices. Here, we have again introduced $\varsigma^0 = i\sigma^y$ and $\vec{\varsigma} = \vec{\sigma}i\sigma^y$ for simplicity of notation.

The total interaction term now has the form
$$\mathcal{H}' = \frac{1}{N} \sum_{\mathbf{k},\mathbf{k}'} [V(\mathbf{k},\mathbf{k}')]^{s_1s_2s_3s_4}_{\alpha\beta\gamma\delta} c^\dagger_{\alpha\mathbf{k}s_1} c^\dagger_{\beta-\mathbf{k}s_2} c_{\gamma-\mathbf{k}'s_3} c_{\delta\mathbf{k}'s_4}, \tag{D.12}$$
where the sums over spin and orbital indices are implied again. The interaction matrix element $[V(\mathbf{k},\mathbf{k}')]^{s_1s_2s_3s_4}_{\alpha\beta\gamma\delta}$ has an odd and an even part in $\mathbf{k}$ which depends on the resulting sign of an interchange of the two first index pairs, $(\alpha\beta, s_1s_2) \leftrightarrow (\beta\alpha, s_2s_1)$. Unlike the case of a primitive unit cell, it is thus not only depending on the spin part. To illustrate this, we look in the following at the specific example of a nearest-neighbor interaction on a square lattice as in section 4.3, i.e., $v(\mathbf{k}-\mathbf{k}') = V[\cos(k_x - k'_x) + \cos(k_y - k'_y)] = -v(\mathbf{k}-\mathbf{k}'+\mathbf{Q})$. The interaction matrix element in this case can be divided into
$$[V(\mathbf{k},\mathbf{k}')]^{s_1s_2s_3s_4}_{\alpha\beta\gamma\delta} = v^+(\mathbf{k},\mathbf{k}')\Lambda^{s_1s_2s_3s_4}_{+,\alpha\beta\gamma\delta} + v^-(\mathbf{k},\mathbf{k}')\Lambda^{s_1s_2s_3s_4}_{-,\alpha\beta\gamma\delta}. \tag{D.13}$$
The two parts are
$$\Lambda^{s_1s_2s_3s_4}_{+,\alpha\beta\gamma\delta} = [(\tau^3)_{\alpha\beta}(\tau^3)^\dagger_{\gamma\delta}](\varsigma^0)_{s_1s_2}(\varsigma^0)^\dagger_{s_3s_4} + [(i\tau^2)_{\alpha\beta}(i\tau^2)^\dagger_{\gamma\delta}](\vec{\varsigma})_{s_1s_2} \cdot (\vec{\varsigma})^\dagger_{s_3s_4} \tag{D.14}$$
with
$$\begin{aligned} v^+(\mathbf{k},\mathbf{k}') &= \frac{1}{2}(v(\mathbf{k}-\mathbf{k}') + v(\mathbf{k}+\mathbf{k}')) \\ &= \frac{V}{2}[(\cos k_x + \cos k_y)(\cos k'_x + \cos k'_y) \\ &\quad + (\cos k_x - \cos k_y)(\cos k'_x - \cos k'_y)] \end{aligned} \tag{D.15}$$
and
$$\Lambda^{s_1s_2s_3s_4}_{-,\alpha\beta\gamma\delta} = [(i\tau^2)_{\alpha\beta}(i\tau^2)^\dagger_{\gamma\delta}](\varsigma^0)_{s_1s_2}(\varsigma^0)^\dagger_{s_3s_4} + [(\tau^3)_{\alpha\beta}(\tau^3)^\dagger_{\gamma\delta}](\vec{\varsigma})_{s_1s_2} \cdot (\vec{\varsigma})^\dagger_{s_3s_4} \tag{D.16}$$
with
$$\begin{aligned} v^-(\mathbf{k},\mathbf{k}') &= \frac{1}{2}(v(\mathbf{k}-\mathbf{k}') - v(\mathbf{k}+\mathbf{k}')) \\ &= V(\sin k_x \sin k'_x + \sin k_y \sin k'_y). \end{aligned} \tag{D.17}$$

As usual, the interaction matrix element decomposes into the so-called extended s-wave channel ($\cos k_x + \cos k_y$) with $A_{1g}$ symmetry, the d-wave channel ($\cos k_x - \cos k_y$) belonging to $B_{1g}$ and the p-wave channel ($\sin k_x$, $\sin k_y$) belonging to $E_u$ in $D_{4h}$. We see, however, that the angular momentum channels of the interaction are not merely depending on the spin channel, but on the combination of spin and orbital channel. Note, however, that this is only true for an inter-sublattice interaction. If we consider an intra-sublattice interaction, such as on-site or next-nearest-neighbor interactions, the interaction is always even in the orbital indices and thus the known dependence of spin and angular channel is restored.

# Appendix E

# Single Layer with Broken Inversion Symmetry

The second example of chapter 4 deals with layers each missing mirror symmetry, but stacked such as to restore global inversion symmetry. For the purpose of comparison, the analysis of the linearized gap equation by Frigeri *et al.* [82] is recapitulated briefly in this appendix. Therefore, we start with a Hamiltonian as derived by these authors,

$$\mathcal{H}^{\text{hop}} = \sum_{\mathbf{k}} \sum_{s,s'} \mathcal{H}^{\text{hop}}_{\mathbf{k}ss'} c^\dagger_{\mathbf{k}s} c_{\mathbf{k}s'} = \sum_{\mathbf{k},s} (\varepsilon_{\mathbf{k}} - \mu) c^\dagger_{\mathbf{k}s} c_{\mathbf{k}s} + \sum_{\mathbf{k}} \sum_{s,s'} (\vec{g}_{\mathbf{k}} \cdot \vec{\sigma}_{ss'}) c^\dagger_{\mathbf{k}s} c_{\mathbf{k}s'}, \quad (\text{E.1})$$

where the first term contains a simple band energy $\varepsilon_{\mathbf{k}}$ respecting both, time reversal and inversion symmetry and the second term is the symmetry-reducing term with $\vec{g}_{\mathbf{k}} = \alpha(\hat{x}\sin k_y - \hat{y}\sin k_x)$. The spectrum of this Hamiltonian is given by

$$\xi_{\pm,\mathbf{k}} = \varepsilon_{\mathbf{k}} - \mu \pm |g_{\mathbf{k}}| \quad (\text{E.2})$$

and we can evaluate the normal-state Green's function to find

$$G_0(\mathbf{k}, \omega_n) = (i\omega_n \sigma^0 + \mathcal{H}^{\text{hop}}_{\mathbf{k}})^{-1} = G_{0+}\sigma^0 - G_{0-}(\hat{g}_{\mathbf{k}} \cdot \vec{\sigma}), \quad (\text{E.3})$$

where

$$G_{0\pm}(\mathbf{k}, \omega_n) = \frac{1}{2}\left(\frac{1}{i\omega_n - \xi_{+,\mathbf{k}}} \pm \frac{1}{i\omega_n - \xi_{-,\mathbf{k}}}\right) \quad (\text{E.4})$$

and $\hat{g}_{\mathbf{k}} = \vec{g}_{\mathbf{k}}/|\vec{g}_{\mathbf{k}}|$. Using this Green's function in the linearized gap equation (4.18) together with an interaction of the form (4.19), we find that, indeed, the singlet and the triplet part of the gap mix,

$$\psi(\mathbf{k}) = -T \sum_{n,\mathbf{k}'} v^+(\mathbf{k},\mathbf{k}') \Big\{ [G_{0+}\tilde{G}_{0+} + G_{0-}\tilde{G}_{0-}]\psi(\mathbf{k}')$$
$$+ [G_{0+}\tilde{G}_{0-} + G_{0-}\tilde{G}_{0+}]\hat{g}_{\mathbf{k}'} \cdot \vec{d}(\mathbf{k}') \Big\} \quad \text{(E.5)}$$

and

$$\vec{d}(\mathbf{k}) = -T \sum_{n,\mathbf{k}'} v^-(\mathbf{k},\mathbf{k}') \Big\{ [G_{0+}\tilde{G}_{0+} + G_{0-}\tilde{G}_{0-}]\vec{d}(\mathbf{k}')$$
$$+ 2G_{0-}\tilde{G}_{0-}\{\hat{g}_{\mathbf{k}'}[\hat{g}_{\mathbf{k}'} \cdot \vec{d}(\mathbf{k}')] - \vec{d}(\mathbf{k}')\} + [G_{0+}\tilde{G}_{0-} + G_{0-}\tilde{G}_{0+}]\hat{g}_{\mathbf{k}'}\psi(\mathbf{k}') \Big\}. \quad \text{(E.6)}$$

Here, we have separated the interaction into even and odd parts again, which we can expand in terms of IRs,

$$v^+(\mathbf{k},\mathbf{k}') = \sum_a v_a \psi_a(\mathbf{k})\psi_a(\mathbf{k}'), \quad \text{(E.7)}$$

$$v^-(\mathbf{k},\mathbf{k}') = \sum_i \sum_b v_b^i d_b^i(\mathbf{k}) d_b^i(\mathbf{k}'). \quad \text{(E.8)}$$

From these two equations, the two main conclusions from the symmetry considerations of section 4.4.1 can already be reproduced. First looking at both equations, we find that for the coupling to the singlet to be maximal, the $d$ vector should lie parallel to $\vec{g}_{\mathbf{k}}$ for all $\mathbf{k}$. For a $\vec{d}(\mathbf{k})$ always perpendicular to $\vec{g}_{\mathbf{k}}$, there is no coupling at all. Additionally taking the $\mathbf{k}$ integration into account, it is obvious that the $(A_{1g})$ singlet indeed only couples to the $(A_{2u})$ triplet. As mentioned in section 4.4.1, the $(D_{4h})$ IR $A_{2u}$ does actually belong to the most symmetric IR $A_{1g}$ in $C_{4v}$. It is of course also possible to construct different singlet and triplet gaps less symmetric where $\vec{d}_{\mathbf{k}}$ is still parallel to $\hat{g}_{\mathbf{k}}$. This is discussed in [87].

Second, when looking at Eq. (E.6) and neglecting the coupling, the linearized gap equation yields

$$\vec{d}_{\|}(\mathbf{k}) = -T \sum_{n,\mathbf{k}'} v^-(\mathbf{k},\mathbf{k}')\{[G_{0+}\tilde{G}_{0+} + G_{0-}\tilde{G}_{0-}]\vec{d}_{\|}(\mathbf{k}') \quad \text{(E.9)}$$

## Single Layer with Broken Inversion Symmetry

for the parallel case and

$$\vec{d}_\perp(\mathbf{k}) = -T \sum_{n,\mathbf{k}'} v^-(\mathbf{k},\mathbf{k}')\{[G_{0+}\tilde{G}_{0+} - G_{0-}\tilde{G}_{0-}]\vec{d}_\perp(\mathbf{k}')\} \quad \text{(E.10)}$$

for the perpendicular case. Evaluating the Matsubara sums in these two expressions [cf. Eqs. (4.46) and (4.47)] and setting $\vec{d}_i(\mathbf{k}) = \Delta_i \hat{d}_i(\mathbf{k})$ with $\hat{d}_i$ normalized we obtain

$$1 = -v_\parallel \sum_{\mathbf{k}'} \sum_{a=\pm} \frac{|\hat{d}_\parallel(\mathbf{k}')|^2}{2\xi_{a,\mathbf{k}'}} \tanh\left(\frac{\xi_{a,\mathbf{k}'}}{2T}\right), \quad \text{(E.11)}$$

$$1 = -v_\perp \sum_{\mathbf{k}'} \sum_{a=\pm} \frac{|\hat{d}_\perp(\mathbf{k}')|^2}{2(\varepsilon_{\mathbf{k}'}-\mu)} \tanh\left(\frac{\xi_{a,\mathbf{k}'}}{2T}\right). \quad \text{(E.12)}$$

For a $\vec{d}(\mathbf{k})$ parallel to $\vec{g}_\mathbf{k}$, there is thus a finite critical temperature due to the divergence of the integral for $T \to 0$. For other gaps with $\vec{d}_\mathbf{k} \perp \vec{g}_\mathbf{k}$, the transition temperature is strongly suppressed for even small antisymmetric SOC, $\alpha \ll t$, since the integral in Eq. (E.12) is not diverging. The suppression of $T_c$ for the case of a finite SOC is shown in Fig. 4.2 (cf. the $t_z = 0$ case).

# Bibliography

[1] Y. Maeno, H. Hashimoto, K. Yoshida, S. Nishizaki, J. G. Bednorz, and F. Lichtenberg, *Superconductivity in a layered perovskite without copper*, Nature **372**, 3 (1994).

[2] R. A. Borzi, S. A. Grigera, J. Farrell, R. S. Perry, S. J. S. Lister, S. L. Lee, D. A. Tennant, Y. Maeno, and A. P. Mackenzie, *Formation of a Nematic Fluid at High Fields in $Sr_3Ru_2O_7$*, Science **315**, 214 (2007).

[3] A. Callaghan, C. W. Moeller, and R. Ward, *Magnetic Interactions in Ternary Ruthenium Oxides*, Inorg. Chem. **5**, 1572 (1966).

[4] G. Cao, S. McCall, M. Shepard, J. E. Crow, and R. P. Guertin, *Thermal, magnetic, and transport properties of single-crystal $Sr_{1-x}Ca_xRuO_3$*, Phys. Rev. B **56**, 321 (1997).

[5] S. Nakatsuji and Y. Maeno, *Quasi-Two-Dimensional Mott Transition System $Ca_{2-x}Sr_xRuO_4$*, Phys. Rev. Lett. **84**, 2666 (2000).

[6] J. F. Karpus, R. Gupta, H. Barath, S. L. Cooper, and G. Cao, *Field-Induced Orbital and Magnetic Phases in $Ca_3Ru_2O_7$*, Phys. Rev. Lett. **93**, 167205 (2004).

[7] M. Kubota, Y. Murakami, M. Mizumaki, H. Ohsumi, N. Ikeda, S. Nakatsuji, H. Fukazawa, and Y. Maeno, *Ferro-Type Orbital State in the Mott Transition System $Ca_{2-x}Sr_xRuO_4$ Studied by the Resonant X-Ray Scattering Interference Technique*, Phys. Rev. Lett. **95**, 026401 (2005).

## BIBLIOGRAPHY

[8] I. Zegkinoglou, J. Strempfer, C. S. Nelson, J. P. Hill, J. Chakhalian, C. Bernhard, J. C. Lang, G. Srajer, H. Fukazawa, S. Nakatsuji, et al., *Orbital Ordering Transition in $Ca_2RuO_4$ Observed with Resonant X-Ray Diffraction*, Phys. Rev. Lett. **95**, 136401 (2005).

[9] J. J. Randall and R. Ward, *The Preparation of Some Ternary Oxides of the Platinum Metals*, J. Am. Chem. Soc. **81**, 2629 (1959).

[10] L. Klein, J. S. Dodge, C. H. Ahn, G. J. Snyder, T. H. Geballe, M. R. Beasley, and A. Kapitulnik, *Anomalous Spin Scattering Effects in the Badly Metallic Itinerant Ferromagnet $SrRuO_3$*, Phys. Rev. Lett. **77**, 2774 (1996).

[11] J. G. Bednorz and K. A. Müller, *Possible high-$T_c$ superconductivity in the BaLaCuO system*, Z. Phys. B **64**, 189 (1986).

[12] A. P. Mackenzie, R. K. W. Haselwimmer, A. W. Tyler, G. G. Lonzarich, Y. Mori, S. Nishizaki, and Y. Maeno, *Extremely Strong Dependence of Superconductivity on Disorder in $Sr_2RuO_4$*, Phys. Rev. Lett. **80**, 161 (1998).

[13] K. Ishida, H. Mukuda, Y. Kitaoka, K. Asayama, Z. Q. Mao, Y. Mori, and Y. Maeno, *Spin-triplet superconductivity in $Sr_2RuO_4$ identified by $^{17}O$ Knight shift*, Nature **396**, 658 (1998).

[14] G. M. Luke, Y. Fudamoto, K. M. Kojima, M. I. Larkin, J. Merrin, B. Nachumi, Y. J. Uemura, Y. Maeno, Z. Q. Mao, Y. Mori, et al., *Time-reversal symmetry-breaking superconductivity in $Sr_2RuO_4$*, Nature **394**, 558 (1998).

[15] J. Xia, Y. Maeno, P. T. Beyersdorf, M. M. Fejer, and A. Kapitulnik, *High Resolution Polar Kerr Effect Measurements of $Sr_2RuO_4$: Evidence for Broken Time-Reversal Symmetry in the Superconducting State*, Phys. Rev. Lett. **97**, 167002 (2006).

[16] T. M. Rice and M. Sigrist, *$Sr_2RuO_4$: an electronic analogue of $^3He$?*, J. Phys.: Condens. Matter **7**, L643 (1995).

[17] S. Das Sarma, C. Nayak, and S. Tewari, *Proposal to stabilize and detect half-quantum vortices in strontium ruthenate thin films: Non-Abelian braiding*

statistics of vortices in a $p_x + ip_y$ superconductor, Phys. Rev. B **73**, 220502 (2006).

[18] F. Lichtenberg, A. Catana, J. Mannhart, and D. G. Schlom, $Sr_2RuO_4$: A metalic substrate for the epitaxial growth of $YBa_2Cu_3O_{7-\delta}$, Appl. Phys. Lett. **60**, 3 (1992).

[19] S.-I. Ikeda, Y. Maeno, S. Nakatsuji, M. Kosaka, and Y. Uwatoko, Ground state in $Sr_3Ru_2O_7$: Fermi liquid close to a ferromagnetic instability, Phys. Rev. B **62**, R6089 (2000).

[20] Y. Maeno, K. Yoshida, H. Hashimoto, S. Nishizaki, S. ichi Ikeda, M. Nohara, T. Fujita, A. P. Mackenzie, N. E. Hussey, J. G. Bednorz, et al., Two-Dimensional Fermi Liquid Behavior of the Superconductor $Sr_2RuO_4$, J. Phys. Soc. Jpn. **66**, 1405 (1997).

[21] D. J. Singh and I. I. Mazin, Electronic structure and magnetism of $Sr_3Ru_2O_7$, Phys. Rev. B **63**, 165101 (2001).

[22] G. Cao, S. McCall, and J. E. Crow, Observation of itinerant ferromagnetism in layered $Sr_3Ru_2O_7$ single crystals, Phys. Rev. B **55**, R672 (1997).

[23] S. A. Grigera, R. S. Perry, A. J. Schofield, M. Chiao, S. R. Julian, G. G. Lonzarich, S. I. Ikeda, Y. Maeno, A. J. Millis, and A. P. Mackenzie, Magnetic Field-Tuned Quantum Criticality in the Metallic Ruthenate $Sr_3Ru_2O_7$, Science **294**, 329 (2001).

[24] A. J. Millis, A. J. Schofield, G. G. Lonzarich, and S. A. Grigera, Metamagnetic Quantum Criticality in Metals, Phys. Rev. Lett. **88**, 217204 (2002).

[25] S. A. Grigera, P. Gegenwart, R. A. Borzi, F. Weickert, A. J. Schofield, R. S. Perry, T. Tayama, T. Sakakibara, Y. Maeno, G. A. Green, et al., Disorder-Sensitive Phase Formation Linked to Metamagnetic Quantum Criticality, Science **306**, 1154 (2004).

[26] P. W. Anderson, Structure of "triplet" superconducting energy gaps, Phys. Rev. B **30**, 4000 (1984).

BIBLIOGRAPHY

[27] T. Giamarchi, *Quantum Physics in One Dimension* (Oxford University Press, 2004).

[28] S.-C. Wang, H.-B. Yang, A. K. P. Sekharan, H. Ding, J. R. Engelbrecht, X. Dai, Z. Wang, A. Kaminski, T. Valla, T. Kidd, et al., *Quasiparticle Line Shape of $Sr_2RuO_4$ and Its Relation to Anisotropic Transport*, Phys. Rev. Lett. **92**, 137002 (2004).

[29] D. B. Gutman and D. L. Maslov, *Anomalous c-Axis Transport in Layered Metals*, Phys. Rev. Lett. **99**, 4 (2007).

[30] A. F. Ho and A. J. Schofield, *c-axis transport in highly anisotropic metals: Role of small polarons*, Phys. Rev. B **71**, 045101 (pages 6) (2005).

[31] M. Turlakov and A. J. Leggett, *Interlayer c-axis transport in the normal state of cuprates*, Phys. Rev. B **63**, 064518 (2001).

[32] N. E. Hussey, A. P. Mackenzie, J. R. Cooper, Y. Maeno, S. Nishizaki, and T. Fujita, *Normal-state magnetoresistance of $Sr_2RuO_4$*, Phys. Rev. B **57**, 5505 (1998).

[33] A. Damascelli, D. H. Lu, K. M. Shen, N. P. Armitage, F. Ronning, D. L. Feng, C. Kim, Z.-X. Shen, T. Kimura, Y. Tokura, et al., *Fermi Surface, Surface States, and Surface Reconstruction in $Sr_2RuO_4$*, Phys. Rev. Lett. **85**, 5194 (2000).

[34] C. Bergemann, S. R. Julian, A. P. Mackenzie, S. NishiZaki, and Y. Maeno, *Detailed Topography of the Fermi Surface of $Sr_2RuO_4$*, Phys. Rev. Lett. **84**, 2662 (2000).

[35] D. Forsythe, S. R. Julian, C. Bergemann, E. Pugh, M. J. Steiner, P. L. Alireza, G. J. McMullan, F. Nakamura, R. K. W. Haselwimmer, I. R. Walker, et al., *Evolution of Fermi-Liquid Interactions in $Sr_2RuO_4$ under Pressure*, Phys. Rev. Lett. **89**, 166402 (2002).

[36] N. Hiraoka, T. Buslaps, V. Honkimäki, T. Nomura, M. Itou, Y. Sakurai, Z. Q. Mao, and Y. Maeno, *Momentum densities, Fermi surfaces, and their*

temperature dependences in $Sr_2RuO_4$ studied by Compton scattering, Phys. Rev. B **74**, 100501 (2006).

[37] T. Oguchi, *Electronic band structure of the superconductor $Sr_2RuO_4$*, Phys. Rev. B **51**, 1385 (1995).

[38] D. J. Singh, *Relationship of $Sr_2RuO_4$ to the superconducting layered cuprates*, Phys. Rev. B **52**, 1358 (1995).

[39] I. I. Mazin and D. J. Singh, *Competitions in Layered Ruthenates: Ferromagnetism versus Antiferromagnetism and Triplet versus Singlet Pairing*, Phys. Rev. Lett. **82**, 4324 (1999).

[40] A. E. Ruckenstein, P. J. Hirschfeld, and J. Appel, *Mean-Field Theory of High-$T_c$ Superconductivity - The superexchange mechanism*, Phys. Rev. B **36**, 857 (1987).

[41] T.-K. Ng, *Spinon-holon binding in t-J model*, Phys. Rev. B **71**, 172509 (2005).

[42] C. Bergemann, A. P. Mackenzie, S. R. Julian, D. Forsythe, and E. Ohmichi, *Quasi-two-dimensional Fermi liquid properties of the unconventional superconductor $Sr_2RuO_4$*, Adv. Phys. **52**, 639 (2003).

[43] A. P. Mackenzie, S. R. Julian, A. J. Diver, G. J. McMullan, M. P. Ray, G. G. Lonzarich, Y. Maeno, S. Nishizaki, and T. Fujita, *Quantum Oscillations in the Layered Perovskite Superconductor $Sr_2RuO_4$*, Phys. Rev. Lett. **76**, 3786 (1996).

[44] M. H. Cohen, L. M. Falicov, and J. C. Phillips, *Superconductive Tunneling*, Phys. Rev. Lett. **8**, 316 (1962).

[45] P. W. Anderson and Z. Zou, *"Normal" Tunneling and "Normal" Transport: Diagnostics for the Resonating-Valence-Bond State*, Phys. Rev. Lett. **60**, 132 (1988).

[46] L. Capogna, E. M. Forgan, S. M. Hayden, A. Wildes, J. A. Duffy, A. P. Mackenzie, R. S. Perry, S. Ikeda, Y. Maeno, and S. P. Brown, *Observation*

of two-dimensional spin fluctuations in the bilayer ruthenate $Sr_3Ru_2O_7$ by inelastic neutron scattering, Phys. Rev. B **67**, 012504 (2003).

[47] R. S. Perry, L. M. Galvin, S. A. Grigera, L. Capogna, A. J. Schofield, A. P. Mackenzie, M. Chiao, S. R. Julian, S. I. Ikeda, S. Nakatsuji, et al., *Metamagnetism and Critical Fluctuaions in High Quality Single Crystals of the Bilayer Ruthenate* $Sr_3Ru_2O_7$, Phys. Rev. Lett. **86**, 4 (2001).

[48] E. P. Wohlfarth and P. Rhodes, *Collective electron metamagnetism*, Philos. Mag. **7**, 1817 (1962).

[49] T. Goto, K. Fukamichi, and H. Yamada, *Itinerant electron metamagnetism and peculiar magnetic properties observed in 3d and 5f intermetallics*, Physica B **300**, 167 (2001).

[50] S. A. Grigera, R. A. Borzi, A. P. Mackenzie, S. R. Julian, R. S. Perry, and Y. Maeno, *Angular dependence of the magnetic susceptibility in the itinerant metamagnet* $Sr_3Ru_2O_7$, Phys. Rev. B **67**, 214427 (2003).

[51] A. W. Rost, R. S. Perry, J.-F. Mercure, A. P. Mackenzie, and S. A. Grigera, *Entropy Landscape of Phase Formation Associated with Quantum Criticality in* $Sr_3Ru_2O_7$, Science **325**, 1360 (2009).

[52] I. J. Pomeranchuk, *On the stability of a Fermi liquid*, JETP **8**, 361 (1958).

[53] H.-Y. Kee and Y. B. Kim, *Itinerant metamagnetism induced by electronic nematic order*, Phys. Rev. B **71**, 184402 (pages 4) (2005).

[54] H. Yamase and A. A. Katanin, *Van Hove Singularity and Spontaneous Fermi Surface Symmetry Breaking in* $Sr_3Ru_2O_7$, J. Phys. Soc. Jpn. **76**, 073706 (2007).

[55] H. Doh, Y. B. Kim, and K. H. Ahn, *Nematic Domains and Resistivity in an Itinerant Metamagnet Coupled to a Lattice*, Phys. Rev. Lett. **98**, 126407 (pages 4) (2007).

[56] A. Tamai, M. P. Allan, J. F. Mercure, W. Meevasana, R. Dunkel, D. H. Lu, R. S. Perry, A. P. Mackenzie, D. J. Singh, Z.-X. Shen, et al., *Fermi Surface*

and van Hove Singularities in the Itinerant Metamagnet $Sr_3Ru_2O_7$, Phys. Rev. Lett. **101**, 026407 (pages 4) (2008).

[57] B. Binz and M. Sigrist, *Metamagnetism of itinerant electrons in multi-layer ruthenates*, Europhys. Lett. **65**, 816 (2004).

[58] K.-K. Ng and M. Sigrist, *Anisotropy of the Spin Susceptibility in the Normal State of $Sr_2RuO_4$*, J. Phys. Soc. Jpn. **69**, 3764 (2000).

[59] S. Raghu, A. Paramekanti, E. A. Kim, R. A. Borzi, S. A. Grigera, A. P. Mackenzie, and S. A. Kivelson, *Microscopic theory of the nematic phase in $Sr_3Ru_2O_7$*, Phys. Rev. B **79**, 214402 (pages 10) (2009).

[60] W.-C. Lee and C. Wu, *Theory of unconventional metamagnetic electron states in orbital band systems*, Phys. Rev. B **80**, 104438 (pages 6) (2009).

[61] H. Shaked, J. D. Jorgensen, O. Chmaissem, S. Ikeda, and Y. Maeno, *Neutron Diffraction Study of the Structural Distortions in $Sr_3Ru_2O_7$*, J. Solid State Chem. **154**, 361 (2000).

[62] R. Kiyanagi, K. Tsuda, N. Aso, H. Kimura, Y. Noda, Y. Yoshida, S.-I. Ikeda, and Y. Uwatoko, *Investigation of the Structure of Single Crystal $Sr_3Ru_2O_7$ by Neutron and Convergent Beam Electron Diffractions*, J. Phys. Soc. Jpn. **73**, 639 (2004).

[63] C. M. Puetter, J. G. Rau, and H.-Y. Kee, *Microscopic route to nematicity in $Sr_3Ru_2O_7$*, Phys. Rev. B **81**, 081105 (2010).

[64] I. Dzyaloshinsky, *A thermodynamic theory of "weak" ferromagnetism of antiferromagnetics*, J. Phys. Chem. Solids **4**, 241 (1958).

[65] T. Moriya, *Anisotropic Superexchange Interaction and Weak Ferromagnetism*, Phys. Rev. **120**, 91 (1960).

[66] W. Metzner, D. Rohe, and S. Andergassen, *Soft Fermi Surfaces and Breakdown of Fermi-Liquid Behavior*, Phys. Rev. Lett. **91**, 066402 (2003).

# BIBLIOGRAPHY

[67] C. Noce and T. Xiang, *A tight-binding model for $Sr_2RuO_4$*, Physica C: Superconductivity **282-287**, 1713 (1997), proceedings of the International Conference on Materials and Mechanisms of Superconductivity High Temperature Superconductors V.

[68] C. Cuoco and C. Noce, *Ruthenates and Ruthano-Cuprate Materials* (Springer Berlin / Heidelberg, 2002), vol. 603 of *Lecture Notes in Physics*, chap. 91.

[69] H. Yamase and H. Kohno, *Instability toward Formation of Quasi-One-Dimensional Fermi Surface in Two-Dimensional t-J Model*, J. Phys. Soc. Jpn. **69**, 2151 (2000).

[70] C. J. Halboth and W. Metzner, *d-Wave Superconductivity and Pomeranchuk Instability in the Two-Dimensional Hubbard Model*, Phys. Rev. Lett. **85**, 5162 (2000).

[71] H. Adachi and M. Sigrist, *Probing the $d_{x^2-y^2}$-wave Pomeranchuk instability by ultrasound*, Phys. Rev. B **80**, 155123 (pages 13) (2009).

[72] H. Yamase, *Effect of magnetic field on spontaneous Fermi-surface symmetry breaking*, Phys. Rev. B **76**, 155117 (pages 11) (2007).

[73] J. Bardeen, L. N. Cooper, and J. R. Schrieffer, *Theory of Superconductivity*, Phys. Rev. **108**, 1175 (1957).

[74] P. Anderson, *Theory of dirty superconductors*, J. Phys. Chem. Solids **11**, 26 (1959).

[75] W. Baltensperger and S. Straessler, *Superconductivity in Antiferromagnets*, Phys. Kondens. Mater. **1**, 20 (1963).

[76] M. Ishikawa, O. Fischer, and J. Muller, *Long Range Magnetic Order in the Superconducting State of Heavy Rare Earth Molybdenum Sulfides and Their Pseudoternary Compounds*, J. Phys. Colloques **39**, C6 (1978).

[77] N. D. Mathur, F. M. Grosche, S. R. Julian, I. R. Walker, D. M. Freye, R. K. W. Haselwimmer, and G. G. Lonzarich, *Magnetically mediated superconductivity in heavy fermion compounds*, Nature **394**, 39 (1998).

[78] Y. Kitaoka, S. Kawasaki, Y. Kawasaki, H. Kotegawa, and T. Mito, *NMR/NQR experiments on heavy-fermion systems and superconductors*, Physica B **359-361**, 341 (2005), proceedings of the International Conference on Strongly Correlated Electron Systems.

[79] G. Knebel, D. Aoki, D. Braithwaite, B. Salce, and J. Flouquet, *Coexistence of antiferromagnetism and superconductivity in CeRh In$_5$ under high pressure and magnetic field*, Phys. Rev. B **74**, 020501 (2006).

[80] S. Vanishri, D. Braithwaite, B. Salce, C. Marin, and J. Flouquet, *Coexistence of bulk superconducting and antiferromagnetic order in the spin-ladder system $Sr_2Ca_{12}Cu_{24}O_{41}$*, Phys. Rev. B **81**, 094511 (2010).

[81] E. Bauer, G. Hilscher, H. Michor, C. Paul, E. W. Scheidt, A. Gribanov, Y. Seropegin, H. Noël, M. Sigrist, and P. Rogl, *Heavy Fermion Superconductivity and Magnetic Order in Noncentrosymmetric $CePt_3Si$*, Phys. Rev. Lett. **92**, 027003 (2004).

[82] P. A. Frigeri, D. F. Agterberg, A. Koga, and M. Sigrist, *Superconductivity without Inversion Symmetry: MnSi versus $CePt_3Si$*, Phys. Rev. Lett. **92**, 097001 (2004).

[83] Y. Yanase, *Random Spin-orbit Coupling in Spin Triplet Superconductors: Stacking Faults in $Sr_2RuO_4$ and $CePt_3Si$*, arXiv:1004.2762 (2010).

[84] K. Machida, K. Nokura, and T. Matsubara, *New Pairing State and Partial Destruction of Pairing in Antiferromagnetic Superconductors*, Phys. Rev. Lett. **44**, 821 (1980).

[85] M. J. Nass, K. Levin, and G. S. Grest, *Bardeen-Cooper-Schrieffer Pairing in Antiferromagnetic Superconductors*, Phys. Rev. Lett. **46**, 614 (1981).

[86] G. Zwicknagl and P. Fulde, *Theory of the upper critical field $H_{c2}$ in antiferromagnetic superconductors*, Z. Phys. B **43**, 23 (1981).

[87] M. Sigrist, in *Lectures on the Physics of Strongly Correlated Systems XIII*, edited by A. Avella and F. Mancini (AIP Conference Proceedings, 2008), pp. 55–97.

Die VDM Verlagsservicegesellschaft sucht für wissenschaftliche Verlage abgeschlossene und herausragende

## Dissertationen, Habilitationen, Diplomarbeiten, Master Theses, Magisterarbeiten usw.

### für die kostenlose Publikation als Fachbuch.

Sie verfügen über eine Arbeit, die hohen inhaltlichen und formalen Ansprüchen genügt, und haben Interesse an einer honorarvergüteten Publikation?

Dann senden Sie bitte erste Informationen über sich und Ihre Arbeit per Email an *info@vdm-vsg.de*.

### Sie erhalten kurzfristig unser Feedback!

VDM Verlagsservicegesellschaft mbH
Dudweiler Landstr. 99
D - 66123 Saarbrücken
www.vdm-vsg.de

Telefon +49 681 3720 174
Fax +49 681 3720 1749

Die VDM Verlagsservicegesellschaft mbH vertritt

MIX
Papier aus verantwortungsvollen Quellen
Paper from responsible sources
FSC® C105338

Printed by Books on Demand GmbH, Norderstedt / Germany